September 2025

Do Well — Do Good!

SUSTAINABLE

SUSTAINABLE

Moving Beyond ESG to Impact Investing

TERRENCE KEELEY

▲ Columbia Business School
Publishing

Columbia University Press
Publishers Since 1893
New York   Chichester, West Sussex
cup.columbia.edu

Library of Congress Cataloging-in-Publication Data
Names: Keeley, Terrence, author.
Title: Sustainable: moving beyond ESG to impact investing / Terrence Keeley.
Description: New York: Columbia University Press, 2022. | Includes index.
Identifiers: LCCN 2022011260 (print) | LCCN 2022011261 (ebook) |
ISBN 9780231206808 (hardback) | ISBN 9780231556668 (ebook)
Subjects: LCSH: Impact investing. | Investments—Moral and ethical aspects. |
Investments—Social aspects. | Sustainable development. | Social responsibility of business.
Classification: LCC HG4515.13 .K44 2022 (print) | LCC HG4515.13 (ebook) |
DDC 174/.4—dc23/eng/20220314
LC record available at https://lccn.loc.gov/2022011260
LC ebook record available at https://lccn.loc.gov/2022011261

Columbia University Press books are printed on permanent and
durable acid-free paper.
Printed in the United States of America

Cover design: Noah Arlow
Cover image: Shutterstock

CONTENTS

To Saskia, who made everything possible . . .
and to Julian and Calum, who made everything purposeful.

FOREWORD

I started my career working in the mortgage market in the 1970s as it was just beginning to develop. In those early days, when mortgage loans were first being securitized into bonds, there were lots of questions—from investors about how to effectively invest in the market, from policy makers about the potential to help promote homeownership, and from others about whether securitizing mortgages was even a good idea.

The mortgage-backed securities market since then has had large ups and downs. Today, however, the market exceeds nearly $10 trillion, and with appropriate underwriting standards plays a vital role in delivering attractive returns for investors while making homeownership more affordable for millions of Americans. Indeed, during more than four decades, the mortgage-backed securities market has saved American homebuyers over a trillion dollars by helping bring down the interest rates on their mortgages.

Having watched the mortgage market grow and evolve in its early days, I see some parallels to the market for sustainable investing today. Embracing these parallels could help avoid excesses like those we witnessed in the 2008 subprime mortgage crisis. Sustainable investing has been one of the fastest growing segments of the asset management industry in recent years, and it remains one of the most frequent topics that we at BlackRock are asked about by many of our clients.

Our job at BlackRock is to offer our clients (who are the actual owners of the assets we manage) a range of choices they can select from to achieve their unique financial objectives. Some choose sustainable investing options, others don't. The choice is theirs. In recent years, the number of clients looking to incorporate sustainability into their portfolios has continued to grow, but, at the same time, so have the critics of sustainable investing. It has sparked a lively debate. It's a debate I welcome. I have always believed that the best way to work through big and important questions is through open and robust debate—challenging conventional ideas, finding new solutions. It is the only way to continually improve outcomes for our clients. In the debate on sustainable investing, it's also critical that we not conflate investment theses with social or environmental objectives.

With *Sustainable*, BlackRock alumnus Terry Keeley has made an ambitious effort to advance today's debate on sustainable investing. Terry brings years of working with clients of all types across the globe, and his book will be relevant to seasoned investors as well as newcomers to the topic. He focuses both on the history of how we have gotten to where we are today and some of the major questions that any honest student of sustainable investing needs to grapple with and help resolve. Over more than two decades of working with Terry, I have always found that he brings unique passion to anything about which he feels strongly. He does that in *Sustainable*—often with humor, sometimes with irreverence.

At BlackRock, our focus is helping clients achieve their investment objectives. In *Sustainable*, Terry tackles even broader themes around the role of business in society and how society can address the most pressing issues facing humanity. In his conclusion, he offers a provocative solution—which he dubs the "1.6% Solution"—for asset owners to consider as they construct their portfolios. I don't agree with all of the opinions or conclusions in Terry's book, but I welcome his contribution, and that of many others, to this critical dialogue.

One issue on which Terry and I agree is the view that reliable data and analytics are going to be critical to the future of sustainable investing. One of the biggest insights I had at the start of my career in the early days of the mortgage market was that investors needed better transparency, data, and analytics to understand the risks and opportunities of the securities they were investing in. That was a cornerstone on which my partners and I founded BlackRock—to bring risk-analytics powered by technology to our clients.

Today, asset owners looking to incorporate sustainability factors into their portfolios have a variety of objectives. Some are seeking to minimize risk, some are trying to maximize returns, and others are seeking non-financial

objectives as well—what Terry characterizes as "doing well and doing good."
The continued evolution of data and analytics in sustainable investment is
critical to helping all of these investors achieve their unique objectives.

At different times during his career at BlackRock, Terry oversaw our official institutions business, ran our educational initiatives, looked after some of
our most important institutional clients, and launched our Financial Inclusion Team, an employee network dedicated to helping those on the margins
of society experience greater financial well-being.

The one constant that Terry displayed time and again across all these roles
was his commitment to helping others achieve their goals. In this book, Terry
carries that commitment forward on an even grander scale.

Larry Fink
Chief Executive Officer
BlackRock

SUSTAINABLE

INTRODUCTION

What if every dollar of global capital has been misallocated, wrongly committed because of outmoded or somehow misguided risk models? This worry seized me after I asked a seemingly innocuous question of a world renowned climate expert at a large investment conference in spring 2021.

"It's 2070, about fifty years from now. Exactly how much warmer are global temperatures?"

Her response was a gut punch.

My climate expert was a key architect of the Paris Agreement in which all but six United Nations member states agreed to enact policies to limit global warming to around 1.5 degrees Celsius versus pre-industrial levels. Better than most, she knew all the daunting tasks that needed to be done for the goals of Paris to be achieved. In short, we human beings must stop pumping carbon into the air and/or find an economical way to take a lot of the carbon that's already there out. She had devoted her entire career to helping policy-makers, corporations, and local communities prepare for the coming energy transition. My job was to interview her before a large audience of investment advisors.

"Well," she started slowly, "the answer is—we just don't know. It could be less than we committed to in the Paris Agreement. But it could also be much more, maybe as high as 4.5 degrees Celsius."

It was a stunning, if honest, answer. It landed on the audience like a neutron bomb. To be clear, an increase of 4.5 degrees Celsius would put us close to the last days of the Paleocene–Eocene era about fifty-six million years ago, when there was no polar ice and seas were at least sixty feet higher. Our world would be unrecognizable.

"Oh my God," we all thought. "What are we supposed to tell our clients?"

It was not a hypothetical question. The job of everyone in the room, myself included, was to advise the world's largest sovereign wealth funds, insurance companies, public pension plans, and endowments about where to invest their money safely for decades to come. The vast majority of their assets were being managed against long-term liabilities—often for retirement needs but also to pay out insurance claims, cover future health care costs, or fund philanthropic endeavors like college scholarships. These assets belong to them and to their beneficiaries. Our job was to help make sure they achieved very specific returns, come hell or high water. It was suddenly clear we could end up with both—both hell *and* high water, that is.

I worried about what was racing through my guest speaker's head, but I was even more concerned about what the audience was thinking. They needed to come away from this interview with actionable insights. Allegedly, I was the expert interviewer. Telling thousands of clients, "Sorry, but the planet is toast, and all your savings may go up in smoke," was simply not an option. My guest had never met most of the folks she was addressing, moreover. I had to find some way out of this abyss.

"Your job," she continued, thankfully, "is to advise your clients about how to prepare for whatever future will come. Uncertainty is inherent in all investing. You need to help them get it right. Of course, you also need to remind them that their investment decisions could impact the outcome."

This longer answer calmed fears somewhat, but it also slipped in an important point: *investors are not innocent bystanders*. Where people ultimately put their money has consequences. Those consequences may make all the difference in the end.

Though certainly top of mind, an existential climate threat is not the only challenge the human family faces in the twenty-first century. There is a daunting list of other geopolitical and social challenges, ranging from Russia's outrageous military adventurism and an ascendent China to growing social unrest caused by mounting disparities between the haves and the have-nots. Equally important, the human family is still recovering from a once-in-a-century body blow—the seemingly never-ending global COVID-19 pandemic. Countries caught in military crossfire and/or lacking

access to vaccines and primary health care could well be scarred for generations. This scarring could hold everyone back. Viruses are a lot like war, after
all; they don't respect national borders. We need to confront our existing
challenges while simultaneously tackling future ones.

Those of us who have been privileged enough to make a living in the
financial services industry have another challenge, too. Increasingly we are
expected to shift from being part of the globe's problems to an important
part of its solutions.

Since the infamous tulip bulb bubble of the early seventeenth century,
there have been recurrent, pointed accusations that the financial services
industry invariably looks after itself while creating imbalance and may-
hem for others. This was a principal conclusion that came out of the Global
Financial Crisis: Wall Street thrived while Main Street writhed. Today, it's
widely thought that the financial services industry actively supports big oil
and thermal coal, intentionally exacerbates income inequality, and is concerned about only one thing: protecting itself and its profits. Some of these
accusations have turned violent. A few years back, hundreds of Occupy Wall
Street protestors were forcibly evicted from tent camps, resulting in three
deaths and many hospitalizations. More recently, protestors have broken into
business offices, including those of my alma mater BlackRock, spilling fake
blood on desks and hallways. Some even glued themselves to headquarter
doors, attempting to force the world's largest asset manager and other financial institutions to abandon their presumed evil ways. Whatever and wherever bad is happening, financiers are assumed to be complicit. It just comes
with the job.

Over the past four decades of my career in finance, I have accepted a
clear, *professional* responsibility. That responsibility is to be a *fiduciary*. To
be a fiduciary means I am responsible for helping my clients—the endowments and foundations, sovereign wealth funds, central bankers, and others
who somehow listen to me—navigate a very uncertain future. More specifically, my job has been to help ensure they and all their beneficiaries achieve
their financial goals within clearly specified and appropriate risk limits. Their
financial interests have always come before my own. I serve by helping them
obtain the financial resources they need to do whatever it is they have to do.

But as a man of conscience, I also have *personal* responsibilities. Well
before I became a fiduciary, I committed myself to living a principle-
centered life inspired by the Catholic traditions that have formed me. My real
"job" is to care for my family and the broader human family while remaining

true to my core convictions. These personal obligations both inform and heighten my professional obligations. Among other things, they require that I examine my actions and words as well as the actions and words of every organization I affiliate with for moral consistency. I have always tried to do so privately and publicly. For example, after the Global Financial Crisis, I wrote about excesses in the financial services industry and examined my *personal* culpability for them.[1] In response, I started a voluntary financial Hippocratic oath movement, dedicating myself alongside thousands of others to abiding strictly by the Golden Rule—that is, treating others only as I, myself, would hope to be treated. I maintain this vow to this day, as a supplement to my other fiduciary commitments. If the asset management industry has been complicit in environmental desecration, selfish profiteering, or any other ignoble pursuit, please know I am wholly prepared to unveil those transgressions and do whatever I can to stop them. After all, my ultimate judge is not of this world. My ultimate goal in this life is to be somehow deserving of the next.

So, now you understand how this book came into being. It is the outgrowth of deep, inextricably interwoven personal and professional responsibilities. In some ways, it has been forty years in the making since that's how long my financial services career has been. It has three immodest yet interrelated goals.

First, it seeks to add clarity and substance to the "stakeholder capitalism" debate. Today, there is a broad consensus that businesses should do more than look after the financial interests of their shareholders. I agree. Businesses have responsibilities to their employees, their suppliers, the communities in which they operate, and our environment, all in addition to their shareholders. A substantial portion of this book examines *how* the corporate world could do more to benefit society and *what* financiers can and should do to promote better corporate outcomes. But here, things quickly get complicated. Businesses, above all else, must create economic vitality. That is their sine qua non. If businesses don't foster economic growth, nothing else will. Without economic growth, moreover, few other forms of progress will prove possible. For its part, finance optimizes risk and reward trade-offs. If finance fails, the whole system fails. Financiers and corporate CEOs are not, nor should they ever be, social or environmental justice warriors, forever prioritizing stakeholders at the long-run expense of their shareholders. This is not their primary responsibility. After a first chapter outlining all that is at stake, I examine this heated stakeholder-versus-shareholder debate, underscoring the urgent need for more comprehensive, longer-term thinking. Most pensioners need their assets thirty years from now, not thirty minutes from now. Our obligations are to them. Fortunately, focusing on a thirty-year horizon transforms many of the decisions we must make today very much for the better.

Second, it seeks to demystify and elucidate the phenomenon of ESG investing, "ESG" standing for Environmental, Social, and Governance objectives. The reason for this is simple: ESG investing is the newest, most sophisticated, and fastest-growing tool the financial services industry has created to help solve the world's most pressing problems. If finance is going to be a crucial part of the solution instead of the problem, ESG investing has got to work. As my chosen profession has birthed it, I am at least inferentially responsible for its success. Of course, this also means I must help ensure it doesn't do more harm than good, a nonacademic worry. All too often, well-intentioned, government-blessed financial innovations grow exponentially before bursting. The proverbial road to hell too often finds good intentions in its pavement. More pertinently, it's possible to agree with all the goals of ESG investing while having some misgivings about all its probable impacts—but now I am getting ahead of myself.

Third, it highlights and drives home the principle of comparative advantage. A fundamental premise of this book is that finance and business have essential roles to play in fashioning more inclusive, sustainable growth, *but they cannot and will not succeed on their own*. If more inclusive, sustainable growth is our ultimate objective, which I deeply believe it should be, business and finance need regulators, public policies, private corporations, civil society, and individuals to play specific, complementary roles. Financiers and businesses have no special powers to right others' wrongs or turn the carbon clock backward. If tax laws, regulatory regimes, and/or personal consumption patterns do not simultaneously support better social and environmental outcomes, business and finance won't be able to compensate for those failings. Comparative advantage helps us understand how to assign the right roles to the appropriate agents. It also helps us imagine comprehensive solutions. A final related and hoped-for bonus of this book is to highlight financial strategies and organizations of unique promise—including green bonds, new forms of impact investing, and many extraordinary nongovernmental organizations (NGOs), which I refer to as *exemplars of hope*. These activities, investments, and organizations can, already are, and will continue to promote more inclusive, sustainable growth. They are essential to achieving our most sought-after long-term goals. I believe there is an optimal way forward—but it's going to require more than our current medley of ESG strategies and endless shaming of publicly listed companies. Much more.

The Nobel Prize–winning and rightly revered professor of economics and finance at Yale, Robert Shiller, has written what is almost certainly my profession's most authoritative book on the role finance can and should play in creating the "good society." Professor Shiller does not go into detail defining what the good society entails. Still, one readily infers that economic efficiency and the Aristotelian conceptions of fairness—that is, fostering social equanimity by correcting all that is inequitable—are essential underpinnings. Shiller's principal thesis is that finance has a specific responsibility for stewarding society's assets efficiently and fairly while simultaneously facilitating its most profound aspirations. A corollary of this claim is that no society can be good—that is to say, no society can be fair *and* efficient—if finance fails to perform its essential assigned duties. In his book, Shiller helpfully profiles many specific skills and sensibilities that more than one dozen types of financial specialists must perform if finance is to help achieve ubiquitous goodness and well-being. These range from mortgage lenders and accountants to derivatives traders. He assigns a notably important role to investment managers:

> Investment managers—those who manage portfolios of shares in companies, bonds, and other investments—are among the most important stewards of our wealth, and thus vitally important players in the service of healthy and prosperous market democracies—all in the service of the Good Society.[2]

Of course, I recognize that many will disagree with Professor Shiller's depiction of investors as vital agents who can, must, and should contribute to social well-being. They likely believe anyone who has anything to do with money forever belongs to the world of Mammon. In some sense, this book is especially directed to you, finance's deepest skeptics. Financiers have given us all plenty to doubt and despise over the years. Believe me, I know! I have witnessed many such failures up close and personally. Note Shiller does not say *all* finance and investment contribute to the realization of the good society, however. Instead, he argues it will not be possible for the good society to flourish unless finance and investment play specific supporting roles. I agree with Shiller. Part of what I hope to do in the pages ahead is to convince you to agree with Shiller, too. Financial professionals and investors have crucial roles to play in maximizing human prospects and achieving optimal environmental, social, and economic outcomes. The principal question addressed in this book is what *specific* roles business and finance should play in helping to create the good society—as well as what everyone else must do so the human experiment succeeds. I think it is possible to create the good society, but we must begin with goodwill and forge a sense of common purpose. We must also be clearer about who should be doing what. I hope you'll agree.

As befits a book of such immodest scope, there is deliberate modularity to its presentation. Certain chapters will appeal to some readers more than others. Chapters 1 to 6 are primarily historical, for example. They set the stage for a more technical deep dive. They help explain how and why stakeholder capitalism and ESG investing have emerged in their current forms and what they are trying to do. They canvass the stakeholder-versus-shareholder debate and dissect the most common arguments waged by modern-day activists. They also review what many CEOs have recently promised to do differently, briefly profile the central role of the United Nations in promoting more responsible investment principles, and lay bare the stultifying financial exigencies of climate change. Laypeople will appreciate this background, but so should financiers. Past is prologue. We won't be able to fashion the world we say we want if we don't fully appreciate all that has led us to where we now are.

Chapters 7 to 11 are the more technical parts of the book. While somewhat harder to navigate, they carefully explain how ESG investing functions today. An understanding of the concept of materiality, how stock market indices are created and updated, governance challenges to "hardwiring" corporate goodness, the dynamic relationship between values and valuations, and a range of indexed and active trading strategies that incorporate ESG themes are essential to understanding what modern ESG investing is and is not, as well as what it can and cannot do. If finance is your field, be sure not to miss these chapters.

The discussion in chapters 12 to 14 gets to the heart of some of the challenges ESG investing faces, including the nonnegligible risk of unintended consequences and its surfeit of verifiable impact. These include the limited efficacy of the "cost-of-capital transmission mechanism," the risk of investment bubbles, the growing probability of unwanted corporate activities shifting from publicly listed corporations to privately or state-owned hands, gross capital misallocation, and the dangerous obfuscation of the roles regulatory and consumer behaviors play. Please don't be intimidated by these concepts; they will all be clearly explained. These chapters also highlight the growing importance of investment stewardship. If ESG is to deliver all that is being asked, the specific perils outlined in chapters 12 to 14 need to be addressed. I hope regulators, financial industry insiders, and corporate CEOs will pay special heed to these concerns. I also hope everyone will come to understand the importance of enlightened stewardship. The best way to improve corporate behavior is from within. And the best way to improve corporate behavior from within is to have the right board and management team working assiduously toward long-term goals that promote all societal interests, including their company's profitability.

The final three chapters of the book focus on solutions. They highlight the importance of civic society and several of the most promising investment techniques that verifiably produce better environmental and social outcomes while also producing market-related returns. They shine a bright light on a number of NGOs that have proven, scalable solutions for promoting more inclusive, sustainable growth; these are my exemplars of hope. These chapters also describe multiple impact investment strategies like green and social impact bonds, which generate investment income while verifiably doing good. It is no use positing a litany of societal and environmental problems without also identifying viable solutions. I bundle all these arguments in a final chapter grandiosely titled "The 1.6 Percent 'Solution,'" with "Solution" appropriately within quotation marks. Let me level-set expectations by explaining what I mean about this right now.

In one fundamental respect, solving our ongoing social and environmental challenges is a question of mobilizing the right amount and types of investment. As it so happens, collectively, humanity has more than enough capital to solve all the problems we choose to address. If we want to deal with excess emissions, economic immobility, polluted oceans, poor public health standards, and rebuilding war-torn lands, we will need to reallocate some assets from their current, less impactful use to more deliberate and productive ends. As it so happens, this is both easier and harder than it sounds. Still, it is doable, and I explain how at the end. In my considered judgment, embracing something like my 1.6 percent "solution"—where a modest share of financial assets is deliberately reallocated to explicit impact investment strategies—would be the best way for modern finance to play its vitally important role in creating the good society. Of course, you could be forgiven for skipping the earlier chapters and jumping straight to these concluding pages to see if they make any sense; that's what I usually do! If I've done my job well, though, my closing chapters might encourage you to go back to the earlier ones just to see what you may have missed. I hope so, anyway.

There is an Irish joke that speaks of a tourist visiting Dublin. He stops a local hurrying by and politely asks how to find one of its most famous pubs, the Temple Bar.

"Oh," the journeyman quips, "I certainly wouldn't start from here." Then, shaking his head despondently, he shuffles away.

Well, we are starting from here. Actually, "here" is the only place we ever get to start from. As you know, "here" is full of problems and worries, including environmental woes we wish we had never created, yet another unnecessary European war, and social challenges we must find some comprehensive way to address. "Here" is not where we want to be.

But "here" is all we've got. And if we do not make the most of "here," we will never get "there." Obviously, "there" is where we need to go.

It's best we get going!

PART 1
The Promise . . .

Chapter One

THE STAKES

Let us not rest until the right livelihood is within reach of every
human being upon this earth.

SWAMI AGNIVESH

We ought to do better for the ones left behind—but I don't think
we should kill the capitalist system in the process.

WARREN BUFFETT

We live in a time for which there is no precedent, though Charles Dickens's
famous depiction of the years leading up to the French Revolution—"It was
the best of times, it was the worst of times . . . it was the spring of hope, it
was the winter of despair"—certainly resonates.

Thousands of academics, activists, and social influencers claim it's the
worst. In their telling, our carefree attitude and "profits matter more than
people" mindset have ruined the environment, relegated millions to poverty, and enriched only a few at the very top. For the first time in history,
most young Americans believe socialism would work better—*socialism*, the
economic system generations of their forebears spilled blood to avoid. For
growing numbers, only one faint hope appears to remain.

*Maybe, just maybe, modern capitalism can undergo the economic equiva-
lent of a deathbed conversion.*

Existential worries relating to climate change lie at the heart of many
of these concerns. On both sides of the Atlantic, millions believe human
flourishing—indeed, the very survival of the planet—depends upon prioritizing the environment over all other stakeholders, especially shareholders. Fancy words, these. Stated plainly, a growing number of folks ranging
from school-age children to their grandparents want corporations to stop
being money-grubbing, short-term-thinking opportunists. Instead, they
insist, businesses should turn their formidable skills and bountiful resources
toward stopping climate change *right now*! While they are at it, many critics

add, businesses should also tackle gender disparities, racial injustice, and ever-widening income gaps. Companies should care as much about their employees, customers, suppliers, communities, and the environment as they do about their profits and shareholders, in other words. Each of these is a stakeholder. Each deserves to be treated with dignity.

And you know, it appears they must be on to something: *a growing number of the same corporate CEOs and financiers accused of wrongdoing agree!*

Capitalism is under fire. If it does not work better for more people or if it continues to spiral unabated toward a climate disaster, evolution could give way to revolution. Government fiats could enjoin free markets. In the most extreme case, humanity could literally self-immolate.

Meanwhile, a related "show me the money" effort is targeting wayward corporations to keep their CEOs' feet to the fire. It's led by an equally engaged army of like-minded *investors* who vow only to buy the stocks and bonds of corporations that deliberately prioritize the needs of their employees, their communities, and the environment. This group is different from the activists and academics, though. Collectively, they oversee *tens of trillions of dollars.* They have more than theories, megaphones, and podcasts: they have *big bucks.* This burgeoning investment practice—called "ESG investing" because of its focus on **e**nvironmental, **s**ocial, and **g**overnance objectives—already includes hundreds of millions of individual investors and thousands of pension plans, sovereign wealth funds, and endowments. ESG investing could well be the biggest thing in finance since the Dutch East India company first issued shares in 1602. No company, country, or market will remain unaffected. ESG's success or failure could literally impact every living creature on Earth.

Obviously, something needs to be done, something big. "Stakeholder capitalism" sounds like it could be the solution, but it's not clear *how* it would be run or *who* exactly would run it. ESG investing also seems alluring. Can ESG investing help save the planet, promote social cohesion, *and* generate great returns all at the same time?

At first glance, it seems that prosperity should and would be more broadly shared and sustainable if, say, corporations focused less on paying higher dividends to wealthy people and more on urgent social priorities. Wealthy folks already live in comfort, and every willing person who desires a living wage deserves an opportunity to earn one. More importantly, humanity can't continue to burn up the earth: there is no planet B. It also seems corporations that expand their priorities from maximizing quarterly earnings to optimizing their technical "ESG scores" would be much better long-term investments.

Right? I mean, isn't all this obvious?

Yes, it's obvious. But sadly, it's not remotely close to being correct.

While it seems incontestable that businesses should shift wherever and whenever they can from being problem creators to problem solvers, this does not mean stakeholder capitalism and ESG investing are sure bets. Wrongly executed, each could fail—and fail spectacularly. That's deeply problematic given all that is at stake. A cursory look at the underlying data shows some of the most strident stakeholder capitalist and ESG claims are easily disproven. For example, many companies do many desirable things—protect the planet, reduce their carbon footprints, take care of their workers, and benefit their communities, all laudatory initiatives that need to be done. But they are simultaneously lousy investments. Being a good company doesn't make you axiomatically "great," not in a financial sense anyway. There is also a host of stakeholder capitalist and ESG claims tossed around that generate deep skepticism. Just ask Professor Brad Cornell of UCLA, who has said, "The ESG concept has been overhyped and oversold. It is backed by weak to nonexistent evidence of promised payoffs for companies, and investors—and is fraught with internal inconsistencies that undercut its credibility."[1] Tariq Fancy, a former head of sustainable investing at BlackRock, has been even more damning: "The financial services industry is duping the American public with its pro-environment, sustainable investing practices. Wall Street is greenwashing the economic system and creating a deadly distraction."[2] In Fancy's telling, finance *is* the problem. No one doubts that ESG is still in its early stages or that much more research and data are needed before definitive conclusions can be reached. But tens of trillions of dollars have already been allocated to products with compelling ESG labels. A great deal now rests upon its success.

Should corporations focus on societal needs at the expense of building superior, desirable products? What algorithm should they use to achieve optimality between growing their market share and paying higher wages? Prioritizing multiple stakeholders raises vexing questions like these. Optimality rests upon assumptions and outcomes that cannot be known in advance. CEOs are no different from the rest of us, after all: they, too, "peer through a glass darkly." Business leaders must make judgment calls with significant trade-offs without the benefit of perfect foresight. They certainly can't predict the future. No one can.

Are firms with smaller carbon footprints better investments than those with bigger ones? When you invest in an ESG-labeled fund, are you making the earth safer and better? Clearly, many investors presume so because that's what they have been implicitly told. I disagree with Tariq Fancy on many things, but he's right about one: *if ESG products are not making the global economy more sustainable and fairer, something else will be needed.*[3] Moreover, there are riches to be made right now by claiming your products are "sustainable" but far too little clarity about what that really means. Yes, some ESG investments serve salient social purposes, and I will identify them. But others lack proven benefits. Worse, overly optimistic financial claims are nearly always the source of market dislocations. Why would this time be any different?

There is ample reason for concern. Current demand for investment strategies that prioritize conscientious companies—that is to say, corporations that do well *and* do good—has never been greater. Nearly $8 billion *per day* has been moved into specific ESG-labeled strategies in the past few years. This is a stunning trend, one that is set to accelerate. It seems only sensible to worry if demand for "sustainable investments" has exceeded the supply of verifiably enlightened corporations with proven social and environmental track records. Investor demand for quality companies that can successfully navigate our pending energy transition—producing the energy and heavy industry products we need while simultaneously lowering the amount of carbon in the air—is especially intense. The implications of this imbalance are obvious but must still be called out: the valuations of many great companies with impressive ESG credentials have been driven to dizzying heights. You don't need to spend any time studying finance to know investment bubbles are harbingers of much more significant troubles. Debilitating spillovers can and often do damage the real economy. Human lives are often ruined in their wake. Market crashes occur when outcomes fail to meet impossibly high expectations.

Equally apparent is an overabundance of academic literature and headline-grabbing books on stakeholder capitalism, all of which proclaim its self-evident, salvific powers. For example, Professors Rebecca Henderson of Harvard Business School and Raj Sisodia of Babson College encourage us to think about *Reimagining Capitalism in a World on Fire* and follow a *Conscious Capital Field Guide.* The latter is a modern-day "how-to" guide for transforming the world one corporation at a time. Another book, *Completing Capitalism*, written by leading executives of the confectionery giant Mars, promises a "healed world" if only businesses first heal themselves. Think "executive exorcism in scale," in other words. A fourth, *Mission Economy* propounds an ambitious "moonshot guide" to changing capitalism by massively restructuring the interrelated roles of governments, corporations, and society, top to bottom. Its brilliant and prolific author, Mariana Mazzucato, analogizes our current situation to how man first successfully landed on the moon. Centralized government planning and massive investment made the inconceivable reality: human footprints on the Mare Tranquillitatis. Regrettably, her proposed solution—that we train and then depend upon the efforts of enlightened bureaucrats to spend, persuade, and ultimately dictate whatever it takes—hasn't worked out terribly well throughout history.[4] Why not? It turns out government officials are no more prescient than those same CEOs I just mentioned, peering through darkened glass. In fact, they are often much less so because they are betting other people's money, not their own. Incentives matter. Misaligning incentives are an all-but-sure recipe for corruption and disaster. Even the eight-hundred-year-old University of Cambridge has weighed in. Their top-ranked press has published a series of handbooks on *The Art of Stakeholder Theory*, instructing global corporations to expand their priorities beyond quarterly earnings, to care more for human hearts by curing societal ills.

Most of this mounting pile of pro–stakeholder capitalism literature is incredibly persuasive. Its broad consensus is that modern capitalism needs to reboot, taking focus away from shareholders and somehow redirecting it toward other stakeholders. On its face, this seems hard to refute. Institutions and advisory firms that have pledged to manage their investments using so-called ESG-integrated strategies collectively managed more than $120 trillion at the end of 2021. With new signatories signing up to the UN's Principles for Responsible Investment every day, this mind-boggling total is projected to exceed $140 trillion by 2025. Given these immense sums, it seems only fair to ask, *Does the globe have upward of $120 trillion worth of conscientious companies and proven ESG strategies to invest in?*

No, it does not. And therein lies a deeply troubling conundrum.

ESG investing is largely driven by idealism and a growing consensus that large money flows can directly change corporate behaviors. Unfortunately, this claim is verifiably untrue. Divestiture does not stop companies from making unwanted decisions; it merely impacts their cost of capital and transfers ownership from those who don't support a given management team and strategic direction to those who more broadly do. Something far more determinant than a higher cost of capital is needed to stop unwanted corporate behavior. The truth is, divestiture is a blunt tool with limited efficacy. This may be why Bill Gates said not long ago, "ESG investing so far has probably not removed one ton of carbon out of the air." Not one ton, even after $120 trillion was ESG integrated? If this statement is true, no further critique of ESG investing is needed; it will have condemned itself. In chapter 14, "Fight or Flee?," you will see why selling the stocks of companies whose behavior you want to change is the last thing you should do. You'll also see how private companies operate under less demanding regimes, which shows there is only so far you can push certain public companies before they privatize. Divestors don't change corporate behavior systemically: only regulators, engaged shareholders, and mindful consumers do.

At its core, this is a financial book. Much of it is devoted to figuring out the possible benefits, perils, and improbable promises of ESG investing. To set the stage for discussions to come, I respectfully ask all ESG investors and those who promote ESG strategies two defining questions: (1) What *verifiable* impacts have your investments had so far on the business world, the real world, and your finances—that is to say, impacts you can truly measure? And (2) If you continue doing exactly what you are doing now, will humanity's most dire environmental and social problems be solved? If these questions are deliberated carefully and answered honestly, I doubt many will be satisfied with the answers.

ESG enthusiasm largely derives from an implicit understanding that ESG investments directly promote environmental sustainability and social progress. Regrettably, this claim is, at best, *partly* true. Some do, and some don't. Investment advisers who identify optimal impact investments and sustainable investment strategies—that is to say, those who ultimately win what I will later describe as "the ESG arms race"—are likely to be well rewarded in this life, and perhaps in whatever life may follow. Conversely, corporate leaders and investment professionals who promise environmental and social advancements they do not or cannot ultimately produce may well face a hellish day of reckoning. They certainly deserve to! This time, though, let's make sure we don't get dragged to hell with them.

Not so long ago, an equally influential group of financial alchemists working fist-in-glove with four presumably prescient, government-accredited agencies—Fannie Mae, Freddie Mac, Moody's, and S&P—took the world through hell and back during the 2008 global financial crisis. Do you recall exactly what happened then? A bunch of good intentions backfired, as they so often do. Allegedly pristine, AAA-rated subprime securities infected our banking systems and caused the most significant financial crisis since the 1930s. Millions lost their jobs, and trillions of dollars of savings were eviscerated. Predictably, those at the bottom got hurt the worst. Like Rumpelstiltskin, a group of technically astute financiers working closely with government-approved housing and credit-rated agencies claimed to have found a magical way to spin straw into gold.[5] Home loans made to so-called NINJAs—those with *no income, no jobs, and no assets* and thus little chance of being able to repay—were repackaged together and sold as risk-free, AAA-rated gold. It ended up being a multi-trillion-dollar hoax! Creating AAA-rated gold out of high-risk, subprime loans was always a crackpot idea. Our banking systems and the economy crumbled tragically when the whole ruse came to light. Are ESG investments like subprime securities? Certainly not. At the heart of the subprime crisis was a bunch of senseless home loans that should never have been made. Given how often history repeats itself, though, we must still ask ourselves, What's to prevent another government-blessed, officially sanctioned investment paradigm—that is, accredited ESG investments—from blowing up in our faces?

The answer: *simple common sense.* The kind of common sense this book hopes to supply in ample measure.

Obviously, the stakes are enormous. Our ultimate goal is to figure out whether and how business and finance can promote human flourishing—fixing what can be fixed, limiting undesirable behaviors, and helping to usher in a new, more inclusive type of market capitalism—while still making sure the global economy continues to expand. To do so, though, we'll first need to agree upon *ends.* Only once we decide where we are going can we proceed to a more fulsome discussion of *means*—that is, how best to get there. I must warn you, though, debates on topics this weighty will get a bit technical before the right solutions can be fairly adjudicated. A discussion of the merits and risks of stakeholder capitalism and ESG investing can proceed productively only on clear, mutually agreed terms. Stakeholder capitalists have plenty of noteworthy critics who can and often do make a robust and contrary case. According to these status quo adherents, the timing and purpose of the heated stakeholder-versus-shareholder debate are falsely premised. Look dispassionately through the long lens of history, they say, and you'll discover all the commotion over creating a better economic system than the one that has dominated over the last half-century is simply impudent. Why fix what's not broken? Staunch defenders of the current system can also back up their rejection of stakeholder capitalist arguments with some compelling statistics. If we're genuinely committed to learning how to create a better economic and financial system—one with the best probability of forging Shiller's good society—it's crucial we first hear the status quo camp out.

Over the past five decades, human progress has been nothing short of breathtaking by nearly every economic and social metric. Boomers have seen real GDP per capita grow by more than five times during their lifetimes, Gen Xers by more than four times. By comparison, per capita income in the first one hundred years of U.S. history spanning four generations—from 1774 to 1860 to be precise—*did not even double.*[6]

Before going any further, in the agreed interests of mutual understanding and respect for the truth, allow me to reiterate the last paragraph. American per capita income grew by four times in the last two generations versus only two times in the country's first four generations. Twice as fast in half the time. Got it?

Now let's ask ourselves the following question: *Does it make any sense to throw out everything that has worked so astonishingly well during the past half-century—generating unprecedented prosperity and wealth—just to experiment with something untried and unproven?* Without question, we have significant challenges, perhaps even existential ones. Undoubtedly, new solutions and approaches are needed. But from a material standpoint, the human family has never had it so good. If we want cleaner air, fewer catastrophic storms, better jobs, more social justice, and better health, we'll have to spend more money, perhaps as much as $275 trillion over the next fifty years, we're told. If so, where will all those trillions come from?

And to be completely clear, it's not about money per se; it's about what money *brings*. Defenders of the status quo can justifiably show how the astonishing wealth creation and economic growth we've experienced over the past three to four decades have axiomatically propelled long-sought progress across many social and quality-of-life factors. Since just 1980, for example, global literacy rates have nearly doubled—from 48 percent to 90 percent— while global life expectancies have increased by more than 26 percent, from fifty-six to seventy-one years. IQ scores simultaneously increased by more than three points per decade, primarily driven by improved diets and better health. Infant mortality has declined tenfold, and access to information via the internet, smartphones, and mobile devices has gone from virtually zero to more than 80 percent of the planet. If we are committed to finding the truth, we should start by recognizing how socioeconomic progress over the past half-century has been simply phenomenal. Activists the globe over are increasingly vocal about making more civic and environmental progress—yet for all their protests, there has never been a better time to be born. And though many of them don't seem to realize it, millennials and Gen Zers are far more blessed than any humans who preceded them. Does it make any sense to tear down a system that has made such abundance possible? Shouldn't we instead analyze what has made all these incredible gains possible and change only what we must to solve our new, evolving priorities?

Defenders of the status quo don't stop there, though. They further highlight how and why social justice advocates have a particular reason to be ebullient. While all income demographics globally, from the richest to the poorest, have improved in the past forty years, those at the very bottom—that is to say, those living in extreme poverty, defined by the World Bank as living on less than $1.91 per day—have improved the most. Not by a little, mind you. On the contrary, those at the very bottom have gained by a lot.

In 1980, nearly half the globe's inhabitants lived in extreme poverty, 46 percent to be precise. By 2020, only about 9 percent did. Of course, 9 percent is still way too high—to be sure, one impoverished person is one too many! Nevertheless, 9 percent is a small fraction of what had persisted for millennia and the lowest ever recorded. To put a human face on this achievement, from 1990 until the COVID-19 pandemic hit, 128,000 fewer people were living in extreme poverty every day.[7] In other words, it took more than one thousand years to raise the first billion human beings out of abject poverty but only the last twenty to liberate the last billion.

Think about this. One thousand years of material human progress—in the last twenty years alone!

It's very important we take complete stock of this. There are reasonable and honorable people on all sides of our current economic and political debates—but each must acknowledge and celebrate this extraordinary phenomenon: *the same capitalism many now decry has nearly vanquished abject poverty.* Billions of human souls who once constantly obsessed about food and shelter of urgent necessity are now able to pursue other improvements in their lives, things like better health care for their children, higher education, and more fulfilling work. This miraculous transformation would never have happened without modern capitalism; in fact, poverty's near eradication is modern capitalism's unique and greatest gift. And while the COVID-19 pandemic tragically interrupted this trend in 2020/21—tens of millions fell back into abject poverty, mainly owing to government-induced economic shutdowns—abject impoverishment can and will fall even further if policy-makers keep global markets open and we remain on course with economic expansion. Indeed, the United Nations now openly muses about the day—perhaps as early as 2030 but no later than 2035—when extreme poverty will essentially be eliminated as a stubborn fact of life, limited only to war-torn countries like Afghanistan and perhaps some poorly governed parts of sub-Saharan Africa. Defeating global poverty was unthinkable only a few decades ago. Ending global poverty now appears almost inevitable. All we must do is keep the global economy growing. Economic growth eradicates poverty, reliably and at scale. That alone makes it invaluable.

Modern rhetoric is routinely reckless. Fanciful theories too often become mainstream. Recent assaults on truth have made many facts unrecognizable. Does this explain why so few today appreciate how ubiquitous growth has eliminated and will likely continue to eliminate poverty at scale? In the recent past, it has done so like no other force in history.

Global populations will not peak until later this century.[8] Continued economic growth remains humanity's best chance to achieve widespread progress against hunger, want, and homelessness. More growth means other crucial social imperatives like better health, broader education, and human fulfillment will also become possible. Sustainable economic growth is especially desirable. If we have more economic growth, we will be able to make more social progress. Those who claim we need to solve our environmental problems by shrinking the economy or that we can cure income inequality merely by redistributing our current, fixed amounts of wealth fail to appreciate the demands growing populations entail. Even if per capita incomes stagnate—something that hasn't happened for decades—rising populations will generate more economic activity. More mouths to feed mean more food must be created. More heads to rest mean more beds and houses must be built. Growth's uninformed critics also fatally underestimate the political challenges austerity invariably brings. If you want to start a revolution, just shrink everyone's paycheck simultaneously. Riots are the voices of the unheard. Anger everyone, and you'll hear from everyone. It would be ugly. Humanity needs continued economic growth to fuel all forms of progress, be it material, social, or—as we will see soon enough—environmental.

But as I mentioned, this is just one side of the debate—the viewpoint of those who defend the status quo. Mostly, it's a backward-looking one. Data is an indispensable part of discerning optimal outcomes and the means of achieving them. The explanatory value of data and its second-, third-, and fourth-order implications must still be debated, however. Stakeholder capital adherents can and often do recognize all the good our current market-based economy has achieved. Stakeholder capitalists are still capitalists, after all! In turn, though, they insist status quo supporters should remain equally open-minded about how and where modern corporations could do more to help humanity. In framing the debate to come, the viewpoint of stakeholder capitalists merits equal time. After all, their goals are indisputably meritorious: *to optimize humanity's prospects in the century ahead, extending economic, social, and spiritual gains to as many members of the human family as possible for as long as possible.* More for all, and more for longer. That sounds good to me. Stakeholder capitalists and enlightened members of the status quo crowd also largely agree on ends; it's only on means where they diverge. Finally, unlike many shareholder capitalists, stakeholder capitalists appear more overtly committed to protecting the planet from growing climate risks. It's hard to see how we will serve humanity's deepest needs if we continue to make our land, air, and water unlivable. So, how might committed stakeholder capitalists make their case?

They start with the obvious: *businesses can be profound forces for social improvement.* Many already are. Given that a great deal of human advancement could be wrought by mindful corporations with almost no cost or effort, stakeholder capitalists say they should try. Consider, for example, all the firms today that are reprioritizing minority-owned businesses in their supply chains; in many ways, they reprise the thousands of delis and other small businesses in southern U.S. states that *finally* pulled down their "whites only" signs in the 1950s and '60s. Ending hateful and discriminatory practices expands economic possibilities to those who need them most—the dispossessed, the forgotten, and the marginalized. Moreover, nondiscriminatory efforts like these also often entail zero or minimal incremental cost. Enacted at scale, eliminating implicit and explicit discriminatory supply chain practices would help many more people with meager means achieve financial well-being. For all these reasons, any businesses that can manage their supply chains more mindfully should. Period.

In addition to supporting more diverse supply chains, stakeholder capitalists claim, there are also many other social problems that businesses and farsighted business leaders are uniquely positioned to tackle. These include diversity in their boardrooms and employee satisfaction and well-being. While achieving these may cost some money, most studies show expenses relating to improved employee satisfaction usually pay for themselves and more through improved productivity.[9] In fact, in most cases worker loyalty is very cheap relative to its dividends. Just look at Costco or Delta Air Lines in the United States or Abcam in the United Kingdom for contemporary evidence.

But like supply chain mindfulness, diversity and employee well-being are also low-hanging fruit. Businesses can promote supply chain diversity, worker loyalty, and greater gender and racial equity in their leadership ranks relatively simply and at negligible net cost. If they can, they should. Period. But what about other, more systemic issues? Aren't there also more complicated and deeply ingrained challenges that imperil human progress that corporations and finance should also help fix?

Yes, there are—the most complex and urgent of which may be climate change.

The raging climate debate summons many emotions. When you filter out the extremes—denials and hysteria—one fact remains beyond contention: human activities, including industrial practices, modern agronomy, and carbon-heavy personal consumption patterns, remain on pace to heighten global temperatures over time. By how much and how fast is abundantly unclear. It is *possible* that temperature rises won't be very meaningful, but it is also possible that the rise could ultimately be as much as 4.5 degrees Celsius versus preindustrial levels. Most reputable science suggests a temperature hike of this magnitude—in other words, a hothouse world—would be catastrophic.[10] Such an increase would make formerly infrequent, cataclysmic weather events the norm. Droughts, wildfires, hurricanes, cold snaps, and heat waves commonly considered once-in-a-century occurrences have seemingly already become common seasonal affairs. Left unchecked, climate change could also upend the lives of billions of people, forcing unprecedented and perhaps even violent migration. Tens of trillions of dollars of property damage and lost economic output would become unavoidable. Tally all this up, and the gross cost and extent of increasingly probable climate-related disasters are mind-boggling. Indeed, if the most extreme projections come to pass, the temporal world to come would not resemble the world we now know.

Given this is a known risk—note I choose the word *risk*, not *certainty*—in this book, at least, climate deniers will receive much the same opprobrium as growth's detractors. To flourish *optimally*, I believe humanity should strive for greener, much less carbon-intensive economic growth. Yes, the global economy needs to continue to grow, but, ideally, it would do so *sustainably*. To not grow sustainably could prove homicidal. To not grow sustainably means we are endangering future generations—that is to say, our kids and grandkids. To not grow sustainably is evidently selfish and shortsighted.

Honestly, who advocates living unsustainably? Whoever they are, they strike me as prime candidates to populate Mars.

A second systemic vulnerability of our times is growing income inequality and its corrosive impact on social cohesion.[11] Even though after-tax and after-transfer income disparity is nowhere near as extreme as pre-tax and pre-transfer income disparity,[12] existing levels of redistribution may still prove inadequate to relieve social tensions and stabilize democracies. In fancy financial parlance, left-hand tail risks for social disruption are not negligible; in lay terms, we are precariously perched atop a social powder keg that could explode with one spark. Without demonstrable progress in the hearts and minds of the broader public, as well as more widespread relief for the genuinely dispossessed, the types of economic protests we've recently witnessed in the United States and around the world could grow more commonplace. The potential for growing violence from disaffected interest groups should be addressed with the deadly seriousness it expressly portends. Violent societies implode from the bottom *in*. Given this mounting concern, note the further progress we've just made in our diagnosis. It is now evident that sustainable growth is insufficient for tackling all the urgent challenges that lie ahead. If prosperity needs to become more widespread than it has been, sustainable economic growth simply won't be enough. What we need is more *inclusive*, sustainable growth—continued economic expansion that mutes the most dangerous impacts of climate change *while simultaneously including* more people who have been or would otherwise be left behind. This tripartite antidote is a more comprehensive solution better suited to redressing our current ills. More growth that helps more people, without over-taxing our air, water, and lands. Frankly, nothing less than more inclusive, sustainable growth may prove sufficient for the complex social and environmental problems that now confront the human family.

In addition to climate change and growing income gaps, a related systemic vulnerability that finance and business could help tackle is *declining economic mobility*. A crucial corollary of economic immobility is wasted human capital. All around the globe, too little human potential is being realized.

Mobilizing untapped human potential could dramatically advance inclusive, sustainable growth. The opportunity here is immense and exciting to consider: hundreds of millions of folks who aren't doing very much could become more gainfully employed and more socially engaged. Mobilizing atrophying human talent sustainably would simultaneously spawn a virtuous, unparalleled boom. Just imagine, if you can, a world where everyone seeking a job with a living wage found one. Growth would multiply *from the bottom up*! Making the global economy more sustainable and inclusive through enlightened employment opportunities and mindful consumption patterns could have multiple aggregating benefits. Every socioeconomic stratum would benefit, most especially the least fortunate. Are business and finance responsible for solving this challenge? No, not on their own, of course. But could they help? Sure, especially in their hiring practices. Stakeholder capitalism appears uniquely well suited to frame and help execute this ambitious plan, too. Stakeholder capitalism strives to account for interdependencies within entire corporate and social ecosystems systematically. It also advocates a heightened focus on intergenerational justice. In other words, stakeholder capitalism explicitly recognizes that decisions we take now will impact generations unborn. With its longer-term orientation, stakeholder capitalism explicitly embraces our obligations to our kids and our kids' kids. Businesses can help by mindfully promoting more economic mobility. If they can, they should.

Today, in most developed economies, the most predictive variable of one's ultimate income level is one's income starting point, a factor readily traced to zip codes. This was not the case in most Western countries just a few decades ago, including the United States. In the 1950s, fifth-quintile American wage earners had a nearly 50 percent chance of getting to the top quintile; today, only 8 percent do. American economic mobility has ossified. Economic ossification has a fatally corrosive social impact: it extinguishes hope. The same is true across much of Europe and other developed nations. Formerly typical "rags to riches" stories have become all but impossible to achieve. The vast majority of those few who now fall into this category are professional athletes, fashionistas, or pop stars. As the Harvard ethicist John Rawls persuasively argued, the idea of anyone being penalized or favored simply because of the circumstances of their birth tears at our innate sense of justice.[13] No one reasonably believes one's starting point in life or one's unchosen familial circumstances should forever be one's ceiling. Individual desires, abilities, and, most importantly, personal initiative and effort should also matter. If not, why proclaim that life, liberty, and the pursuit of happiness are inalienable rights?

A fourth systemic social issue that has suffered from insufficient progress over the decades is long-standing civil inequities relating to race and gender. How much business and finance can do about these is unclear, but they are stubborn problems that should be fixed. American justice isn't color blind. Other judicial systems are equally bad or worse. All around the globe, local jurisprudence and civic policing discriminate against those considered socially or racially inferior. George Floyd's shocking, senseless murder brought routine, thuggish police brutality far too long confined to America's shadows into everyone's homes and consciences. For millions, myself included, seeing George Floyd suffocate to death under the knee of an indifferent white police officer was a societal call to arms. Let George Floyd's indefensible murder by a police official be the last. Rife racial profiling and routine hate crimes that we continue to witness from Minnesota to Myanmar and Marseilles are abhorrent. They must end. Their perpetuation is a catastrophic failure of our collective conscience and communal responsibility. We are, each of us, called to rise above such hatred. Our aspirations should be animated by the principles of human dignity and universal solidarity. Can corporations be a vital part of the remedy? Perhaps, at least in part. If so, they should be. Stakeholder capitalism offers a viable path for bringing business on board as part of a broadscale effort to redress all injustice. At the very least, business and finance must not be bystanders to injustice.

As with race, the extreme lack of female leadership in political and corporate halls of power can be explained only by bias, not merit. Today, 8 percent of Fortune 500 companies are led by women. Does anyone honestly believe that only one out of every twelve CEOs deserves to be female in today's age? If not, become part of the solution. Every female CEO who overcame dispiriting odds and reached the top has an inspiring tale of struggle to tell, like Indra Nooyi of PepsiCo, or Katharine Graham of the *Washington Post*, the first woman to head a major U.S. firm.[14] Inspirational stories like these should be trumpeted and replicated. These remarkable women are the luminaries in whose footsteps our daughters and granddaughters can now follow.

Absent more deliberate efforts to include every human being in our economic and social ladders of opportunity, with heightened and enlightened attention to race and gender, carcinogenic inequities will persist. More inclusive, sustainable growth is a helpful mission statement for systemic progress. Inclusivity has multiple dimensions across race, gender, and social class. If finance and business can be a part of this transformation, they should be. Inclusivity and sustainability are what stakeholder capitalism is all about.

In addition to these four broad categories of societal and environmental challenges, it is distressingly easy to identify other urgent societal needs. Each seems unconscionable when a privileged few have so much. Indeed, in a world where there are more than three thousand billionaires and U.S. corporations alone generate more than $3 trillion of profits per year, why is it that:

- Three in ten people still lack access to safe drinking water
- 650 million youth lack basic math and literacy skills
- Less than half the global population receives essential health services
- Eight hundred million students have no access to online learning
- Two billion adults are excluded from formal financial services
- 40 percent of countries have fewer than ten doctors per ten thousand people

This list could go on. Is it the responsibility of business and finance to help solve all these and every other problem I have not listed? In one sense, yes— *but only up to the point that the next few chapters make abundantly clear.* If we are to achieve broader human flourishing, the global economy must be vibrant, labor markets must be fair and robust, and capital must flow efficiently and more purposefully. As a complement to these factors, however, public policies, civic society, and individuals must be equally committed to forging more inclusive, sustainable outcomes. Unless every stakeholder in society does their part, all the challenges I have painstakingly enumerated here cannot and will not be solved. In the most human terms, they are the stakes. There is a mind-boggling litany of unfinished work and new challenges that all men and women of conscience—be they stakeholder or shareholder acolytes, status quo adherents, conscientious citizens, or passionate activists—justifiably want addressed in the months and years to come. Our collective hunger for solutions and earnest desire for more inclusive, sustainable prosperity define the challenges of our time. For this reason, *more inclusive, sustainable growth* will be the North Star that illuminates every page of this book. As we evaluate corporate policies, public programs, civic organizations, and investment products, we will repeatedly ask, Do they help or hinder the tripartite goal of more inclusive, sustainable economic growth? And when we see obstacles to inclusive, sustainable growth we will ask, How best can they be mitigated or removed?

All of us—individual citizens, policy-makers, businesses, public and private institutions, financiers, civic organizations, and regulators—have the opportunity and responsibility to coalesce around a deliberate, ambitious plan for more inclusive, sustainable growth. Moreover, we should address all that is at stake with a heightened sense of urgency; further delay means more wasted human capital and unnecessary loss of life, as well as a genuine risk of portions of the planet becoming unlivable. More inclusive, sustainable growth is needed, and more inclusive, sustainable growth is needed *right now*.

So, now that the stakes are clear and our desired *ends* broadly agreed, we will turn our attention to *means* and methodologies. What's the best way to promote more inclusive, sustainable growth? Should we use some "stakeholder" or "shareholder" model, whatever these may mean? Will ESG investing help or hinder us on this urgent journey? What must each of us do that is different and accretive to get where we have just agreed we need to go? Finally, what are the unique responsibilities of business and finance?

It is appropriate to turn to John Mackey as we begin to answer these questions. John Mackey is the inspirational founder of Whole Foods and, in many ways, the modern human embodiment of stakeholder capitalism. He is also a central figure in the next chapter. Mackey recently summed up our predicament and prospects crisply:

> Free-enterprise capitalism is the most powerful system for social cooperation and human progress ever conceived. It is one of the most compelling ideas we humans have ever had. But that does not mean we should not aspire to something even greater.

In my opinion, John Mackey is right. In the twenty-first century, the human family has both the opportunity and responsibility to aspire to and achieve something more significant—that is, more wholesome living standards, more human fulfillment, broader social justice, more inclusive growth, and a sustainable environment. We have achieved so much, but we should still lift our gaze even higher. Free-enterprise capitalism has undoubtedly been a gift. We should celebrate all it has brought us. But it must now be optimized anew and redirected to address a new list of unique challenges, the challenges of our time.[15] Progress is neither automatic nor inevitable. It must be forged. To be forged successfully, shared values and wise vision must be translated into viable action plans. Failure to achieve consensus on these three fronts— values, vision, and viable action plans—will ultimately translate into lost human welfare. All progress depends upon improving human livelihoods.

So, what must we do now? First, we will begin envisioning an action plan for more ubiquitous well-being by determining which form of capitalism is best suited to facilitate more inclusive, sustainable growth: one oriented toward serving shareholders or one that attempts to enlarge its lens by committing to serving a broader class of stakeholders. Only after untangling the heated stakeholder-versus-shareholder debate can we proceed to a deeper discussion of what ESG investing is, where it came from, how it works, and what it's capable of doing and not doing. And once we fully plumb the promise and perils of ESG investing, we will turn our attention to what more business, finance, and other economic agents need to do to fashion a future that best fulfills human aspiration. Soon enough, we will be able to comprehend the potential relevance of what I call the "1.6 percent 'solution.'" In this plan, we learn where to find the private capital needed to solve our most urgent problems, as well as how best to allocate it.

Chapter Two

STAKEHOLDERS VERSUS SHAREHOLDERS

I would like to put a nail in the coffin of the ideology of
Milton Friedman. His ideas will asphyxiate our democracy and
ultimately be the end of us all.

DARREN WALKER, PRESIDENT, FORD FOUNDATION

Shareholder value is the dumbest idea in the world.

JACK WELCH, MANAGER OF THE CENTURY

Henry Ford is America's first and foremost example of a stakeholder capital-ist in more ways than one. He thought all people could and should live their lives in ways never imagined—traveling by motorcar rather than by foot, hoof, or bike. As a result, he revolutionized daily life, dramatically improved the lives of his workers, and greatly benefited the communities in which he and his factories operated. He also made himself and his fellow shareholders fabulously wealthy in the process.

Ford's big breakthrough, of course, was the Model T (figure 2.1). Collo-quially known as the "Tin Lizzie" or "Leaping Lena," the Model T was the first car mass-produced that middle-class Americans could readily afford. Days after its first release on August 12, 1908, an astonishing fifteen thou-sand orders were placed. By 1927, fifteen million had been sold. In 1999, the Global Automotive Elections Foundation named the Model T "Car of the Twentieth Century." Elon Musk can only dream of similar accolades for the twenty-first.

"I will build a car for the great multitudes," Henry Ford promised. "It will be constructed of the best materials by the best men to be hired—but so low in price that no man making a good salary will be unable to own one."

FIGURE 2.1. Henry Ford with his "Tin Lizzie," 1921.
Wikimedia Commons.

BULL'S-EYE!

By 1916, the Ford Motor Company had accumulated a capital surplus of $60 million—the equivalent of more than $1.6 billion today. Over the previous eight years, the price of Ford's most important product, the Model T, had been reduced to *half* its original starting tag, and the wages of Ford's workers had simultaneously grown more than *sixfold*. Convinced what he was doing was good, Ford sensibly concluded that more would be even better. He announced a plan to use his accumulated capital to build more plants and expand production. His publicly stated ambition was one of the earliest and clearest cases ever made for stakeholder capitalism:

> My ambition is to employ still more men, to spread the benefits of this industrial system to the greatest number possible, to help them build up their lives and their homes. To do this, we must put the greatest share of our profits back in the business.

But Henry Ford had one problem: he did not unilaterally control his company. His innate desire to serve a broad class of stakeholders at the short-term expense of his fellow investors was hotly contested by two minority shareholders: John Francis and Horace Elgin Dodge. John and Horace were brothers and, in the fullness of time, direct competitors of Ford's. The Dodge brothers argued it made no sense to keep reducing prices when Ford could barely keep up with the orders still coming in. They sued the company and insisted that Ford's board of directors continue to pay them, as shareholders, special dividends from its capital surpluses rather than build more plants or cut profit margins.

Remarkably, Michigan's Supreme Court ruled in their favor on February 7, 1919. Not only did this decision repudiate Henry Ford's stakeholder instincts and desires, but it also helped codify into law as well as common practice the primary interests of shareholders, a standard that has been upheld in varying degrees implicitly and explicitly ever since. According to Justice Ostrander's majority opinion,

> A business corporation is organized and carried on primarily for the profit of the stockholders. The powers of the directors must be employed for that end. The discretion of directors is to be exercised in the choice of means to attain that end but does not extend to a change in the end itself—i.e., to the reduction of profits or the nondistribution of profits among stockholders.[1]

In the context of the current stakeholder-versus-shareholder debate, it is hard to overstate the importance of Ostrander's decision. A century ago, in the most unambiguous terms possible, a compelling case for stakeholder capitalism was struck down by Michigan's highest court, and shareholders were unambiguously awarded the upper hand. In effect, the legal case for shareholder supremacy had been settled. But Henry Ford was not a man to be dissuaded nor a man to alter his ambitions because of a few itinerant partners. He ultimately bought out the Dodge brothers and continued to expand his plants and dominion. Through most of its storied history, too, the Ford Motor Company never lost focus on its workers' quality of life or the communities in which they operated. In 1926, the Ford Motor Company became the first large U.S. firm to announce two days a week off for its workers, a radical idea at the time. A few years later, two days a week off became the U.S. corporate norm. Henry Ford also hired many African Americans and women long before most Americans considered diversity socially desirable.[2] At least through its first quarter-century, the strength of Ford's underlying business model allowed his company to become ever more generous with opportunities, wages, and other benefits, generating an effective cycle of virtue. And by his death, Henry Ford had amassed the equivalent of a $200 billion fortune—depending on where Amazon and Tesla are trading today, close to what Jeff Bezos or Elon Musk may have.

It is tempting to think that the preoccupations and concerns of one's time are unique and that there is little to be learned from the experiences of decades past. But the current discussion on how best to enlarge and improve the social contributions of business traces its roots back centuries, well before Henry Ford's day. Before the Enlightenment, multiple stakeholders—including the Crown, Guild, Church, and landowners—extorted rents from peasant workers. Multiple constituencies taxed their businesses, making them far less efficient than they could and should have been. Royal charters were granted as early as the thirteenth century. The earliest in England established the University of Cambridge. The first U.S. corporations were granted licenses to build and operate public works, providing communities with clean water, constructing and operating transnational railroads, and creating mutual-benefit insurance companies. As corporate formations expanded under different states' laws before the Civil War, businesses grew in strength and economic impact, almost always without any direction or official rules. For this reason, the stakeholder-versus-shareholder dispute remained largely dormant during the first half of the nineteenth century. Businesses were seen as essential contributors to communities. Cities and states cheered their arrivals and successes. They facilitated vital societal and commercial needs. Communities depend upon successful, organized commerce after all. The modern debate about the social purpose of business was forced to center stage only late in the nineteenth century, after transgressions by prominent actors in the railroad and oil sectors violated the public's trust. In the 1880s, apparent abuses by railroad and oil monopolies, just like later failures by financial firms during the Great Depression of the 1930s and the Great Recession of 2008/09, sparked intense national conversations about how corporations should behave and what they owe to society, beyond their commercial contributions and the enrichment of their shareholders.

One of the most illuminating of those conversations occurred in 1931 and 1932 between the law professors E. Merrick Dodd of Harvard and Adolf A. Berle of Columbia. In his 1931 *Harvard Law Review* article, "Corporate Powers as Powers in Trust," Professor Berle argued that the role of a corporate board trustee was to maximize the interests of shareholders as the sole beneficiaries of the corporation. In essence, Berle sided with Justice Russell Ostrander and Michigan's Supreme Court. One year later, in the same journal, Professor Dodd countered that corporations must have both profit-making and social-benefit functions, as the right to incorporation is effectively granted by society. Dodd's point was that the absence of any social role would effectively preclude the need for and right of incorporation. Berle's subsequent response directly bears upon the challenges faced by boards and ESG investors today. Though society may well desire that corporations do more for their communities, shareholder primacy de facto must prevail because so-called stakeholder needs lack clear metrics for setting and measuring success. As he wrote,

> A social-economic absolutism of corporate administrators—even if entirely benevolent—might be unsafe, and in any case hardly affords the soundest base on which to construct economic industrialism which the commonwealth seems to require.[3]

In other words, Berle is saying that even though some critics may want businesses to do more good, their sine qua non is to promote economic vitality. Commonwealths require economic industrialism. Corporations in their first and final incarnations are economic agents. They should not be hamstrung or set up for failure. The most tangible thing they can contribute to society is being successful. The most tangible way to measure their success and viability is by their profitability. What being "good" means forever lies in the eye of the beholder, meanwhile. Like beauty, its definition is subjective and open to endless debate. What is seen as good or beautiful or politically correct today may not be considered the same tomorrow. What is both tangible and necessary should not be sacrificed for the illusory.

The Dodge versus Ford Motor Company story is especially worth telling in this context. It illustrates the centrality of shareholder claims in both the strategy and success of a publicly listed corporation. As it so happens, shareholders not only have a unique primacy in *academic* management decision theory; they also have a *legal* basis for that primacy. Like it or not, corporate profits cannot be handed out willy-nilly to make society fairer or greener or more just without implicit or explicit shareholder consent. Shareholders could sue a corporation for, say, overpaying their workers or giving away their earnings to a charity. If they did, most importantly, they would probably win. The Michigan Supreme Court case and the famous Berle–Dodd debate also debunk one of the most common myths stakeholder capitalists routinely espouse today: that the Nobel Prize–winning economist Milton Friedman somehow singlehandedly gave rise to the concept of shareholder value. Obviously, Professor Friedman did no such thing.

As the Michigan Supreme Court case and the Berle–Dodd debate show, the concept of shareholder rights has significant social and legal precedent. But the University of Chicago's Milton Friedman still somehow felt an urgent need to give it renewed impetus in the fall of 1970. It was then, in the most quoted article that the perpetually spry Professor Friedman ever wrote, that shareholder interests were seemingly deemed above all others:

> There is one and only one social responsibility of business—to use its resources and engage in activities designed to increase its profits so long as it stays within the rules of the game.

It's often said that historians are more powerful than God because they can change the past, something God, even with infinite powers, cannot do. Accusing Professor Friedman of starting capitalism's obsession with shareholder value illustrates this witticism. When Professor Friedman's article "The Social Responsibility of Business" appeared in the *New York Times* on September 17, 1970, no one batted an eyelash, let alone protested in the street. In many ways, Professor Friedman's assertions then seemed the epitome of common sense. Yet those who ardently defend stakeholder capitalism today—including Darren Walker of the Ford Foundation[4]—often cite Friedman's article as the opening contretemps for their cause.

As tempting as it may seem to some, Professor Friedman did not just wake up one morning determined to reject all that Henry Ford had popularized and what Professor Dodd had argued so folks like the Dodge brothers could become ever wealthier. Not at all. Much more urgent matters weighed on his mind. Failing to appreciate those urgent considerations not only does a disservice to Professor Friedman's intellect, ethics, and argument, but also heightens the risk that those extenuating problems could once again repeat themselves. Milton Friedman excelled at common sense. As it so happens, the specific circumstances that beset Professor Friedman then also bear heavily on the stakeholder-versus-shareholder debate today. In a nutshell, economic growth should always be considered a top priority. Without it, few other societal goals can be reached.

Unlike all other Allied and Axis powers, the United States emerged from World War II with its industrial capacity largely intact. As might be expected of a valorous victor, the United States took active steps to maintain its edge but also to assist its prior foes—Japan and Germany—to recover. With its leverage and comparative advantage, the U.S. economy first experienced consistent nominal growth in high single digits, most especially in the 1950s. By the 1960s, however, America's formidable economic engine began to sputter. Competition from the very same economies the United States had helped rebuild began to bite. During this emergent slowdown, Professor Friedman felt it necessary to warn his compatriots that U.S. businesses were losing their edge. Inflation was north of 5 percent and rising; GDP was barely registering 1 percent and was falling. Unemployment was stubbornly stuck above 6 percent, meaning labor markets had failed to assimilate hundreds of thousands of vets returning home from the Vietnam War. Scant few of those vets were even welcome to return to their homes and communities, meanwhile. Many of their fellow citizens saw them as traitors to American ideals. Four months earlier, marking a nadir in U.S. social history, thirteen students at Kent State University had been gunned down by Ohio National Guardsmen as the students peacefully protested the expansion of American military action into Cambodia. Earlier that same day, Kent State students had symbolically buried a copy of the Constitution, claiming it had been murdered because Congress had not authorized the attack on Cambodia. President Richard Nixon had sent bombers and troops into a sovereign state all on his own. Trust in the federal government plummeted to an all-time low. Nixon had, after all, pledged to end the Vietnam War, not expand it. Four of those thirteen students died. While many presume social unrest in America today is the worst ever, the late sixties and early seventies were arguably much worse. Protesting students were being shot and killed. The U.S. economy was sclerotic.

Professor Friedman's mounting concerns about corporate profits at that time were entirely warranted. The U.S. federal government had massively increased spending on domestic priorities and the military while twenty-eight of the thirty companies composing what was then the Dow Jones Industrial Average (DJIA) were experiencing declining margins and slowing sales. If the economic engine that had powered U.S. prosperity fitfully during the 1950s and '60s were to stall out further, as it then very much seemed it might, addressing any U.S. threat, foreign or domestic, would become simply untenable. Power abroad stems first and foremost from power at home. No country can progress without domestic economic vitality.

Milton Friedman was no evil dope. And history went on to prove his 1970 concerns were entirely prophetic. As figure 2.2 shows, between 1968 and 1980, the DJIA had a *negative* return, something that had happened only once before over a comparable period. Yet this lost economic decade was not the result of any burst bubble, as in 1929 or during the senseless dot-com dalliance in the late 1990s. Instead, it occurred because corporate America had lost its capacity to innovate and grow. Corporate America in the late sixties and seventies failed at its most crucial task: *economic vitality*. Corporate America failed its citizens in the 1970s by not generating sufficient prosperity. Professor Friedman intentionally wrote in 1970 that corporations needed to focus more on profitability at the expense of all other variables because he foresaw corporate America was developing a chronic profitability problem. U.S. prosperity was fading away. America was running out of geese capable of laying golden eggs.

It has long been understood by academicians and practitioners alike that capitalism has implicit strengths and weaknesses. Unbridled capitalism maximizes efficiencies while degrading certain societal assets, like the environment or job security. None of this is new. These social costs are referred to in economic textbooks as "negative externalities." Stakeholder capitalism aims to maximize capitalism's strengths while lowering or ideally eliminating all

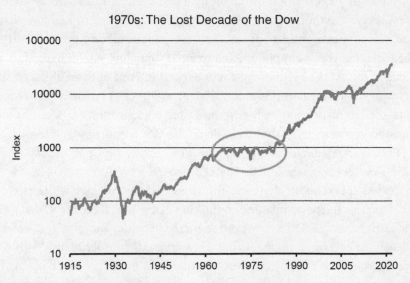

FIGURE 2.2. The lost decade of the Dow: U.S. stocks went nowhere from 1968 to 1980.
Figure by the author.

negative externalities. In addition, stakeholder capital enthusiasts want businesses to generate targeted *positive* externalities. Positive externalities are things like cleaner water and air, improved local schools, and public parks. Indeed, for the most ardent stakeholder capitalists, the more positive externalities corporations generate, the better.

But capitalism itself comes in multiple forms. There is no monolithic definition of capitalism. Getting our "capitalist terminology" straight is essential if we hope to delineate *good* capitalism from *bad* capitalism. Agreeing how and where capitalism can be good and bad is also essential to understanding the stakeholder-versus-shareholder debate.[5]

Most references to capitalism obliquely refer to what should instead be called *free-market* capitalism. This version has been most commonly found in the Anglo-Saxon world, at least until recently. It contrasts with *state-guided* capitalism, which has been raised to a new art form today in China. Both free-market and state-guided capitalism generally allow markets to create consumer goods and services and set prices freely. They primarily differ in that governments drive the investment agenda in the latter. China's domestically lauded "Made in China 2025" initiative is a recent, relatively extreme example of government-planned investing. It set out a roadmap for China to dominate multiple technology growth areas for decades to come. China's Communist Party chiefs actively planned to become global market leaders in artificial intelligence, electronic vehicles, semiconductors, renewable energy, and all things 5G through effectively limitless government investment. Unlike those of the West, China's economy relies much more upon state-directed investments than consumption or private investments. By investing heavily in tomorrow's technology, Chinese leaders planned to circumvent the all too common and unfortunate *middle-income trap*. A middle-income trap occurs when emerging-market economies can't reach peak growth because workers' wages can't be further increased; exports would then lose their competitiveness. More skilled, higher-tech industries are needed to break the logjam.

Given that the government runs its economy centrally and issues its own currency, China's ability to invest in higher-tech industries is effectively unlimited; whatever capital they need will be found. Today, most Western adaptations of capitalism are hybrids of free-market and state-guided versions. Still, China is state-guided capitalism's undisputed hegemon.

A third type of capitalism—*oligarchic capitalism*—commonly grows out of state-guided capitalist constructs. Oligarchic capitalism shares many of socialism's least praiseworthy characteristics, including high levels of corruption, chronic inefficiency, and substandard growth.[6] Russia has been oligarchic capitalism's undisputed hegemon, at least up until their tragic invasion of the Ukraine. Other notable examples include Argentina and Belarus. Finally, there is *entrepreneurial capitalism*, a fourth category. Entrepreneurial capitalism scores highest in generating growth and prosperity. No society can consistently achieve better living standards without some element of entrepreneurialism. Macroeconomic dynamism also relies on micro dynamism: just as no economy can grow optimally without entrepreneurialism, no corporation can long dominate without retaining some entrepreneurial spirit. Entrepreneurialism's hallmark is never-ending innovation. Entrepreneurialism is needed to breed geese that ultimately lay the golden eggs. Without entrepreneurialism, economies are relegated to less growth, sluggish productivity, and compromised standards of living.

But entrepreneurialism, too, has a turbulent side. Its underlying disruptive force is the heart of what the Austrian economist Joseph Schumpeter famously called capitalism's "creative destruction" tendency. As Schumpeter memorably described it, "a gale of creative destruction drives the process of industrial mutation that continuously revolutionizes the economic structure from within, incessantly destroying the old, incessantly creating the new." In Schumpeter's powerful depiction, capitalism's positive and negative externalities are opposite sides of the same coin. Established businesses (think taxicab or livery car drivers) are often undermined by technical advances and human ingenuity that seek greater efficiencies (think, for the sake of this example, Uber). As the old gets washed away by the new, it's invariably uncomfortable for those in the disrupted industry. In the end, however, a more efficient match between consumer preferences and corporate service emerges. Over time, as this new, more efficient equilibrium emerges, greater productivity and, ultimately, more prosperity result. Trying to stop these "creative destruction" innovations is often a fool's errand, moreover; too much regulation and you merely end up with sclerotic, oligarchic outcomes. From a pure efficiency standpoint, it's often preferable to accept short-term discomfort for longer-term gain. Failing to do so means other, more productive methods, companies, and societies may ultimately pass you. One other point must be made about entrepreneurialism, though: its financial riches are almost always concentrated among a favored few. In this respect, entrepreneurial capitalism reprises Winston Churchill's observation about democracy: just as Churchill claimed democracy was the worst form of government except for all the others, entrepreneurialism is the worst form of capitalism except for all the others. Every economy needs some form of entrepreneurialism for its perpetual renewal. But even perpetual renewal comes with a price. Absent progressive tax codes to spread resulting wealth more broadly or worker retraining to facilitate greater labor force mobility, rampant entrepreneurialism puts social contracts under pressure. We are witnessing such social contract pressures in America and China today. President Xi is calling for "more common prosperity," and America's perennial tax debate is about making the rich pay their "fair share."

Returning to Friedman, the DJIA composition in 1970 usefully illustrates how nonentrepreneurial the U.S. economy had become. About half the thirty corporations in the DJIA mined minerals, smelted metals, and produced chemicals, including three that exclusively drilled for and refined oil. Six produced autos, auto parts, aircraft, or other durable goods, while four others—General Mills, Sears, Woolworth, and General Foods—primarily catered to U.S. consumer needs. Eastman Kodak, Westinghouse Electric, and AT&T were the closest equivalents to what we would now consider "tech" or growth. But, no early-stage Apple, Microsoft, or Salesforce was included. In fact, only one of 1970's DJIA 30 today remains in the Dow: Procter & Gamble. In other words, the DJIA of 1970 depicts a U.S. economy entering an era of ossification.[7] Without more Schumpeterian creative destruction, a lost American decade was inevitable. Criticizing Friedman today is no different from calling out Isaiah or Jeremiah in the eighth or seventh century BCE—both prophets in their day who were spurned because they warned troubled times were coming. And just like Isaiah and Jeremiah, Friedman also proved correct.

While Ford Motor was never part of the DJIA, its fortunes, too, were ultimately whittled down in time by Schumpeterian forces. Domestic and global competition driven by more farsighted innovation as well as a comparatively high-cost base reduced Ford's share of the global car market from its peak of nearly 60 percent in the early part of the last century to a little over 5 percent recently, half that of the current global automobile leader, Toyota. Ford's recent stock market capitalization of approximately $70 billion also reflects investor expectations of its relative growth prospects versus, say, Tesla, which is sixteen times larger. The markets believe the future of the automobile market is electric and that Tesla is poised to win that race. Ford remains an essentially family-run business with a good labor record; indeed, their recently announced electric car factory will be staffed only by union workers. For decades, the Ford Motor Company was what stakeholder capitalists—and some ESG investors, for that matter—dreamed of: a company that did much good for its employees, community, suppliers, *and* sharehold-ers. Nevertheless, failing to stay at the forefront of the creative-destruction automotive entrepreneurial race has made Ford an unattractive stock to own for decades. Ford's stock price hit a high of $56.68 on June 30, 1998, but lan-guished mainly between $10 and $15 a share from 2002 until recently. From 1998 to 2021, Ford's stock fell—80 percent—while the S&P 500 rose more than 400 percent. It's been a dog.

The founder and CEO of Whole Foods, John Mackey, occupies a place as equally storied as Henry Ford in the debate of stakeholder-versus-share-holder capitalism, but not for the reasons you might think. Like Ford, Whole Foods revolutionized how many people lived their lives, pioneering a highly successful supermarket concept for natural and organic foods. Millions who could not access high-quality, locally sourced organic foods in the United States, Canada, the United Kingdom, and Mexico found their sublimated desires sated when one of more than five hundred Whole Foods Markets opened in or near their neighborhoods. Just like Ford, Mackey also started his company with $45,000 raised mainly from family and friends. Recog-nizing his pioneering spirit, the accountancy firm Ernst & Young named John Mackey "Entrepreneur of the Year" in 2003. And when Amazon bought Whole Foods for $13.7 billion in June 2017, Mackey and his fellow sharehold-ers pocketed significant financial rewards.

But in addition to all this, and much more pertinently for this discussion, John Mackey has perhaps gone to more lengths than any other modern CEO to promote the stakeholder capitalist movement. Along with coauthor Raj Sisodia, he wrote the bestselling book *Conscious Capitalism: Liberating the Heroic Spirit of Business*. And most importantly for our current discussion, in 2005, Mackey debated a still spry, ninety-three-year-old Milton Fried-man on stakeholder-versus-shareholder capitalism. That debate provided an opportunity for Professor Friedman to enunciate how his understanding of shareholder capitalism was always intended to be stakeholder focused. As you review this debate, you should decide for yourself how much of the stakeholder-versus-shareholder debate is semantic. Is their charged argu-ment born more of subtle verbal distinctions than substantive policy dif-ferences? More importantly, do Friedman's actual, rather than purported, beliefs best align with your own?

John Mackey began his telling 2005 exchange with Professor Friedman by laying out how Whole Foods painstakingly monitored the value it created for six categories of stakeholders: customers, team members (his term for employees), investors, vendors, communities, and the environment. Mackey then told an illustrative story about Whole Foods' frequent practice of "5 percent days." Several times a year, on well-publicized dates, Whole Foods announced that 5 percent of a given franchise's total sales would be donated to a local nonprofit, charity, or NGO that was doing good work. With margins well above 5 percent, Whole Foods' beneficence still left plenty of room for profits. But note how these charities were often chosen with a dual profit motive behind them. Mackey revealed that priority was often given to local groups with large memberships who were also infrequent Whole Foods customers. Bringing in new or lapsed customers who appreciated how 5 percent of their expenditures would support the cause they held most dear generated immediate brand loyalty for Whole Foods. From a commercial standpoint, this loyalty was not dissimilar to the loyalty Henry Ford generated by making his cars affordable to those who were assembling them or investing in local communities where his plants were operating. Indeed, it was the quintessential win–win, doing well while doing good. Whole Foods bred brand loyalty and top-line revenues while simultaneously practicing calculated altruism.

John Mackey, after all, built a successful business. Whole Foods' peak profitability exceeded hundreds of millions of dollars. Unsurprisingly, Professor Friedman wasted no time in celebrating Mackey's self-serving virtuosity, quoting one of Adam Smith's most memorable lines from *The Wealth of Nations*: "By pursuing his own interest, an individual frequently promotes that of the society more effectually than when he readily intends to promote it." Nonagenarian Friedman then summarized his complimentary opinion of the objectives of stakeholder capitalism by citing his original 1970 *New York Times* article in his defense:

> The differences between John Mackey and me regarding the social responsibility of business are, for the most part, rhetorical. Strip off all the camouflage, and it turns out we are in essential agreement. Moreover, his company Whole Foods behaves in accordance with all the principles I spelled out in my 1970 *New York Times* article. It could hardly be otherwise, as his company has done well in a highly competitive industry. Had he devoted a much more significant share of his resources to social responsibility unrelated to his bottom line, Whole Foods would be out of business by now or taken over. Some profit had to be made.
>
> With all its defects, the current, largely free-market and private-property-based world seems to me vastly preferable to an alternative world in which very large amounts of resources would be distributed to 501c(3)s by their corporate counterparts.

In other words, Professor Friedman is saying let businesses be businesses and charitable organizations be charitable organizations—that is, don't violate the law of comparative advantage. If working with a charity is good for business, go ahead. If not, stay away. And, of course, it was here that Professor Friedman would have liked to end the debate. But John Mackey would have none of his equivalencies. Mackey reconfigured his argument by hypothesizing about the *causal direction* of profits. After all the dust surrounding the virtues of stakeholder-versus-shareholder capitalism settles, it may be that we are left with nothing more than the age-old debate: which comes first, the chicken or the egg? Or, in this case, *does profitability enable virtue, or does virtue enable profit?* Mackey said,

> Professor Friedman is right to argue that profit-making is intrinsically valuable for society. But while he believes taking care of customers and employees and the environment and philanthropy are means to the end of increasing investor profits, I take the exact opposite view: making high profits is the means to fulfilling Whole Foods' core business mission. We want to improve the health and well-being of everyone on the planet through higher-quality foods and better nutrition. We cannot fulfill this mission unless we are highly profitable.

Mackey justifiably argues that Whole Foods' core mission is noble: health is essential to human flourishing, and organic foods help promote health. But let's be brutally honest: how scalable is Whole Foods' positive symbiosis? It is easy to imagine many businesses for which profitability and virtue are inextricably intertwined, with mutual dependence beyond doubt—companies like Disney or Humana, for example. Whole Foods benefited commercially from being perceived as healthy and altruistic. But not all businesses have this same symbiotic opportunity. Indeed, some consumer brands have risen to global prominence through explicitly nonbenevolent imaging. For example, the deliberate cultivation of an image of vice or illicit purpose has benefited firms like Victoria's Secret and Remington. And many others—Sony, IKEA, Coca-Cola, and Rolex—have achieved strong brand identities based upon attributes well beyond virtue or vice. Privilege, exclusivity, and simplicity come to mind. The world's most valuable brands—Apple, Google, Amazon, and Microsoft—achieved their distinct identities first and foremost through outstanding products and services, not through altruism. Each may have added some altruism on the edges, but superior phones, tablets, search engines, delivery services, and software were and still are the primary source of their success. Without desired products, they would have lacked commercial impetus. If Mackey had picked pipe welding or truck driving as his core business instead of organic food, would he have been able to make the same causal case between virtue and profits? Certainly not. Yet doesn't the world still need pipe welders and truck drivers? Yes, it very much does.

Another crucial question arises from the Friedman–Mackey debate: Is it possible to discern whether the commitment to serve one's stakeholders in the pursuit of higher profits is any different from setting out to run one's company in the best way possible for one's shareholders *over long periods*? Here we have briefly considered auto manufacturing and grocery stores. In the cases of Ford and Whole Foods, tending to their multiple stakeholders helped both succeed for prolonged periods, though as we've seen, Ford eventually lost focus on competitiveness and innovation. In many other industries, circumstances also seem broadly the same: stakeholder interests and shareholder interests align more consistently when considered over longer time horizons.

Consulting firm McKinsey's Global Institute examined this precise question in some detail. After reviewing the strategies and stock performance of 615 large- and mid-cap companies from 2001 to 2015, they found exactly what we just intuited: *companies with a longer-term planning horizon significantly outperform their short-term-oriented peers across a wide range of economic and financial metrics.*[8] McKinsey's data shows that firms that invest more in their employees, manage margin growth more consistently, focus less on quarterly earnings, and support their core stakeholders (especially through difficult times) deliver considerably higher returns on average to their shareholders. Earnings growth among the long-term-oriented exemplars in this McKinsey study was 36 percent higher than for the short-termers. In fact, nearly half of all earnings growth was captured by corporations with longer-term behavioral characteristics in their sample, even though they represented about one-quarter of the sample by number. If long-term profit maximization is indeed the goal, prioritizing the interests of all one's stakeholders—that is to say, one's employees, customers, communities, and the environment—appears ultimately to inure to the benefit of shareholders as well. Stated more succinctly, *over the longer run, stakeholder capitalism and shareholder capitalism appear broadly synonymous.* Companies that fixate on short-term profits at the expense of the variables that feed their long-term growth don't live long to tell their tales. That said, as evidenced by Ford's longer-term shortcomings, inattention to shareholder interests ultimately fails all other stakeholders as well. Every corporation must keep an eye out for Schumpeter. Every company must constantly innovate and compete to remain relevant and serve all its stakeholders optimally.

Shareholders don't win over time when companies cut corners, fail to invest in future growth markets, or abuse their employees. Yet all large-scale, publicly traded businesses must ultimately be profitable to impact society in the manners most desired by those who ascribe to stakeholder capitalism. When they aren't profitable, they perish. Note this does *not* mean every corporation *must* be profit obsessed, however.[9] Many important and impactful smaller businesses can and often do operate effectively as mutual-benefit societies or as nonprofit corporations. It is also possible to incorporate and be publicly traded yet not seek profits as one's penultimate goal. After all, this appears to have been Ford's implicit strategy. However, what is less clear is how many shares of these so-called good companies belong in your retirement portfolio. Owning the stocks of supremely virtuous but unprofitable public companies is unlikely to translate into personal retirement security. Equally illustrative, large-scale, *private* businesses can and often do endure vicissitudes of earnings lapses for prolonged periods as they look after other investment needs and stakeholder priorities.[10] But large private businesses cannot avoid the exacting demands of turning a profit forever, either. There are fewer than six thousand public companies in the United States and well more than five million private corporations. Given public companies in America employ fewer than 20 percent of working Americans, the more extensive discussion relating to inclusive, sustainable growth that we will soon unpack goes well beyond the reach of public shareholders. Any sensible manifesto for maximizing inclusive, sustainable growth must extend beyond the domains of public corporations as well.

All stakeholders matter, including shareholders. Profitability is the sine qua non of all business success that is not intended to be charitable—public and private. Business success is a necessary—but by itself insufficient— condition for human flourishing. While such claims are likely to draw derisive responses from those who have not operated within the business world, they remain beyond practical contention. For humanity to flourish, businesses must drive our economic success. If they are not generating economic vitality, communal prosperity and social cohesion will suffer. A society without economic prosperity will struggle to find the resources to do the things it most needs and wants. A society without economic prosperity cannot achieve social, political, or even environmental well-being.

But as I discussed in the opening chapter, this is not how many activists now see the world and its woes. If Greta Thunberg were to have her way, much more transatlantic travel would be by sailboat. If Reclaim Finance were to have their way, nary one more drop of oil would leave the ground. And if Engine No. 1 were able to wave a magic wand, the boards of many corporations would be fundamentally transformed. Each believes that heightened urgency on our pending climate transition and other critical social needs should be immediately brought to the fore.

In fact, many activists believe there is no way for humanity to progress without uprooting the entire system—shuttering all oil companies, abandoning free markets, and forcibly redistributing wealth. Their passions infuse many elements of our current public policy debates. It is to their arguments and passions we now turn.

Chapter Three

ACTIVISTS, THEIR ARGUMENTS—
AND A LITTLE ENGINE
THAT COULD

If there is no struggle, there is no progress. Power concedes
nothing without a demand; it never did, and it never will.
FREDERICK DOUGLASS

Grown-ups have failed us.
GRETA THUNBERG

"Corrupt, deceitful, and dangerous to their core."

That's how the Australian-based activist group StopAdani describes the Indian energy and industrial conglomerate that is the singular focus of their movement. StopAdani's website details a litany of Adani Group's alleged transgressions, from the illicit use of tax havens, fraud allegations, corruption investigations, and human rights abuses to sunken coal ships, pollution, secrecy, and bribery. If any are accurate, we should all be outraged. Stop-Adani's unrelenting criticism has made them one of the most effective global campaigns to starve an individual company of their political, social, and financial capital.

Unsurprisingly, StopAdani's accusations stand in stark contrast with the content of Adani's website. Adani's corporate motto, "Growth with Goodness," dominates their home page, and a prominent drop-down "Sustainability" tab enumerates dozens of environmental, social, and community milestones. While greenwashing on websites and annual reports is standard operating procedure for many industrial and energy corporations, it is especially egregious among the biggest scofflaws. Lots of smiling faces, lush landscapes, and blue skies commonly mask environmental horrors underneath. In Adani's case, it's not just their website that praises their efforts, though: in 2017, they won an award from CSRWorks for best sustainability reporting practices in Asia.

A necessary and passionate argument rages about how ethical Adani Group is or isn't. But this debate should not be conflated with how heavily the world's second most populous nation relies upon their infrastructure, logistics, defense, aerospace, agribusiness, and—most importantly—energy production capabilities. Adani's multiple coal thermal power plants combined with a large and growing number of solar and wind power projects make them India's largest private electricity supplier, by far. Without Adani's thermal coal power plants, tens of millions of Indians in Gujarat, Maharashtra, Rajasthan, Karnataka, and Chhattisgarh would have no electricity. State-owned power plants would need to step in. Unfortunately for India and the planet, though, nearly all of India's public plants are coal fired as well. State-owned oil and energy companies are often among the most egregious environmental offenders, but government agencies seldom experience the same scrutiny as publicly traded corporations. Even when they are scrutinized, they are often slow to respond.

There is also no doubt thermal coal ranks at the top of the most environmentally destructive electrical power sources. Coal's greenhouse gas emissions, dirty mining practices, and waste are second to no other power source in the world in terms of carbon effluents and environmental degradation. Yet thermal coal remains the number one electricity source across much of Asia—not just in India but also in China and most nations in the Association of Southeast Asian Nations (ASEAN). Coal's reliability, affordability, abundance, navigable technology, and safety records have made it the preferred electricity source in most developing countries for decades. Coal power can be produced at any time (not just when it is sunny or windy) and is a relatively inexpensive fuel source of which the globe has more than three hundred years of proven supply. These advantages and their historical context help explain why 40 percent of the globe's current electrical production still comes from coal—and in India's case, why coal generates more than 70 percent of the country's current electrical output. Despite recent advances in renewables, they also help explain why many Asian countries intend to build more coal-fired plants, especially where resources like wind and sun are less reliable. About 80 percent of the world's future coal-powered electrical plants are planned in five Asian nations alone: China (187 gigawatts), India (60 gigawatts), Vietnam (24 gigawatts), Indonesia (23 gigawatts), and Japan (9 gigawatts). This means Asian economies are poised to generate much more carbon in the years immediately ahead, not less.

Carbon Tracker is a nonprofit financial think tank based in the United Kingdom whose research promotes more sustainable energy markets through sophisticated, climate-aware capital market analysis. They contend investors should steer clear of all these coal projects. Once you factor in the declining costs of renewable energy and the likelihood that governments and consumers in the future will insist upon cleaner energy to achieve their goals, Carbon Tracker finds more than 90 percent of Asia's coal-based units will be uneconomic.[1] Moreover, their analysis suggests that coal-fired plants will become economically obsolete far more quickly than their proponents contend. This means investors could lose hundreds of billions of dollars by commissioning new ones. "The last bastions of coal power are swimming against the tide when renewables offer cheaper solutions that align with global climate targets," Catharina Hillenbrand von der Neyen, a senior research analyst at Carbon Tracker, pointedly observes. "Investors should steer clear of new coal projects, many of which are likely to generate negative returns from the outset."

Hillenbrand's is a compelling argument. It also contrasts with StopAdani's in meaningful ways. StopAdani is trying to starve a single publicly traded firm of capital by publicly shaming them and their operating business model. If Adani can't find anyone to buy their debt or insure their projects, StopAdani seems to reason, Adani will have no choice but to change course. Carbon Tracker chooses to wage a different fight. They acknowledge and accept the legal rights of public corporations to source private capital to maximize their returns. But they advise the potential providers of that capital to take heed given that projected returns from coal projects are not as lucrative as they may seem. Rather than engaging in a moral argument against one company, in other words, Carbon Tracker provides financial analysis that should dissuade possible funders of Adani and all other coal enthusiasts on *economic* grounds. Investors should avoid funding coal plants because of underappreciated and often overlooked financial risks. In addition to environmental principles, Carbon Tracker appeals to pocketbooks.

Reclaim Finance—an NGO and think tank based in Paris—apparently concurs with StopAdani's approach. Like StopAdani, Reclaim Finance believes starving errant public companies of capital will force them to change their behaviors. Whereas StopAdani looks first to a transgressing corporation, though, Reclaim Finance begins with asset managers and banks. The primary focus of their animus is capital *providers*, not capital *takers*. Of the two, Reclaim Finance is also by far the most ambitious; rather than target one or several companies, they target the entire free-market system. Their goal is wholescale reform, in the vein of what the Occupy Wall Street movement (figure 3.1) sought to do.

In multiple publications, Reclaim Finance has enumerated a long list of their financial service "demands." First, banks, insurers, investors, central banks, and other stakeholders "must end all direct support for new or existing coals mines, power plants, and infrastructure projects." At the same time, divestment and the complete preclusion of all financial services to any enterprise that engages in oil and gas exploration, new pipeline construction, or the building of terminals for the import and export of liquefied natural gas must take place immediately. Ambitiously, Reclaim Finance further insists that financial institutions "sanction companies that sell rather than close their carbon assets." Finally, they demand that financial institutions require their corporate clients to "[align] their activities with the 1.5°C target" and "integrate workers' rights and local community rights in business transformation and train them [workers] for sustainable jobs in the future." Reclaim Finance promises to implement strategies "of increasing pressure, to be deployed if engagement fails or delivers poor results." In short, for Reclaim Finance, the buck stops with those who invest the bucks, not with those who receive them. Reclaim Finance insists that banks, insurance companies, and all other financial institutions cut off their money lines to corporations that won't immediately align with Paris or explain why they cannot. If the lending and investment institutions do not comply, Reclaim Finance attacks them, impugning their motives, condemning their actions, and tarring them as enablers.

It's a remarkably plucky strategy. For Reclaim Finance, the key to achieving more inclusive, sustainable growth lies in making sure banks, insurance companies, asset managers, and financiers stop facilitating anything they consider environmentally unsound or socially unacceptable. Evil will cease to exist when and only when financiers take all money away from evildoers. If bankers and insurers and asset managers don't divest and shun, Reclaim Finance considers them complicit. It does not matter if—as is the case with indexed asset managers—the allegedly transgressing institutions neither own the assets nor make the underlying investment decision; their clients do, through their selection of certain investments. For Reclaim Finance, any entity that allows its clients to invest in firms they don't like will be showered with scorn. It doesn't even matter if those clients can buy the same index from dozens of other purveyors and assume the same exposures through simple derivative transactions. Nor does it matter who makes up the indices.

FIGURE 3.1. Occupy Wall Street May Day demonstration.
Monika Graff / Stringer, Getty Images.

Of course, the idea of starving abhorrent practices of financial resources—be they corporate or personal—is hardly new. For centuries the practice of making money from money—that is to say, the charging of interest on loans, also known as usury—was strictly disallowed by most Jewish, Islamic, and Christian traditions. "Thou shalt not lend upon interest to thy brother," reads Deuteronomy 23:20–21, the fifth oldest book of the Torah, written sometime between the seventh and fifth centuries BCE. For many devout Muslims, lending with interest is still forbidden—hence the prevalence of Sharia-compliant financial products in broadly Islamic jurisdictions. The Catholic Church still proscribes *excessive* interest charges through c2354 of the *Code of Canon Law*. Yet, these extensive religious laws against usury, with all their authority and reach, did not end lending with interest. Far from it. Through the centuries, while many faithful complied, interest-bearing loans were always available. Nonbelievers and noncompliant believers continued to lend with interest, filling the gap. Ironically, those lenders also made more money than they would have without usury laws since they faced reduced competition.

Much more recently—in the past few decades, to be precise—an exclusionary financial practice has been applied most aggressively to so-called sin stocks: companies involved in socially-questionable activities, often in industries like tobacco, gambling, alcohol, and firearms. This practice occurred under the broad rubric of "socially responsible investing." The primary way sin stocks were penalized was through their removal from investment portfolios and benchmarks. However, as with religious rulings against usury, sin stock exclusion had virtually no impact on individual corporate behaviors. Sin stock excluders followed their beliefs and enjoyed peace of mind in so doing—but gambling, alcohol and tobacco use, and firearm purchases continued unabated, as did their production. For the targeted, investment exclusion meant nothing to their core business practices.

As with usury, the most significant practical challenge for those seeking to starve certain businesses of funding is that it takes only a small number of noncompliant individuals and institutions to undermine the strategy. Social activists seem impervious to one commonsense principle of finance: *adequate funding is invariably found when the underlying commercial activity it supports is broadly legal and generates financially acceptable returns.* Yes, widespread exclusions *may* force capital costs to go up, but they then reach a new equilibrium. That new equilibrium usually means *higher* returns to the remaining investors moreover, not lower returns. That, after all, is the converse of "higher cost of capital": more reward for the provider of that capital in the form of higher dividends or interest payments. This is what happened to most companies in designated sin industries, in any case. While some people stopped owning the shares of transgressing industries, others were quite happy to own more. And, of course, the share prices of Philip Morris, Wynn Resorts, Anheuser-Busch, Smith & Wesson, and their industrial brethren never went to zero. Nor did any of these "sinners" ever abandon their core business. The shares of leading tobacco, gambling, alcohol, and firearm companies merely transferred hands from those who refused to own them to those who were less conflicted. Nor was this simply a matter of someone else being *less* ethical, mind you. In most jurisdictions, regulations effectively require investors to own shares like sin stocks through so-called fiduciary rules about which I spoke in the introduction. Fiduciary rules effectively promote objective risk–reward decision-making. Fiduciary standards require financial intermediaries to put their clients' financial interests first, independent of their values or any other considerations that may impede superior returns for that matter. This means financial-exclusion campaigns create investment opportunities that others are often required to exploit. Targeted businesses were forced to reexamine their commercial activities, of course. This led to a greater focus on their core constituents and customers.

Ironically, though, the end effect for many targeted firms was more consistent and higher, more qualitative earnings.

Academic studies evaluating returns on baskets of sin stocks versus socially responsible firms also reveal it is often better to be "bad" than "good," at least from a financial perspective. "If you are looking to make a solid investment, moral convictions aside, a diversified portfolio that includes both saints and sinners is often the better choice," concludes Lisa Smith.[2] And, of course, gambling, guns, and intoxicants are not the only "sin investments" one can cite. An even more appalling, multi-billion-dollar business—pornography—continues to find ample investment capital because of one coldhearted reason: strong consumer demand for the underlying product.[3] If public naming and shaming haven't starved the porn industry of capital, one should wonder what commercial activity shaming could stop.

Taking this back to our opening example, if consumers continue to demand reliable and cheap electricity, and if coal-fired plants provide that power better than alternatives—both fairly reasonable bets across much of the world today—firms like Adani and governments like those of India and China will find the capital to produce it no matter what activists say. Attacking the supply side of unclean energy has some impact, to be sure: it drives up the cost of capital. But higher capital costs alone cannot and will not eliminate the building of more coal plants. If you want to change such behaviors, the regulatory environment and demand dynamics for electricity must be impacted as well.

Remarkably, following this argument to an extreme only strengthens it. Even *outlawing* unwanted businesses can and often does fail to impact demand meaningfully enough to alter their funding. In America's Prohibition era from 1920 to 1933, a nationwide constitutional ban on the production, importation, transportation, and sale of alcoholic beverages did almost nothing to end American demand for booze. Rather than eliminate the production and delivery of intoxicants, in fact, Prohibition perversely made alcohol a more lucrative business by driving it underground. Organized crime and petty criminal gangs pocketed millions by bootlegging beer and liquor supplies. For the public, Prohibition became a double whammy: their communities lost valuable excise tax revenues while the costs of policing skyrocketed. Ultimately, the rejection of Prohibition was more popular than its imposition. The Eighteenth Amendment to the Constitution establishing Prohibition was overturned by the passage of the Twenty-First Amendment. Consumer demand trumped all other considerations in the end. It's rumored

FDR sipped a dirty martini—his favorite, including a splash of olive brine—immediately after signing the Twenty-First Amendment into law.

Professors Brad Cornell of UCLA and Aswath Damodaran of NYU used the same lines of the argument discussed earlier in a recent investigation of highly ranked ESG firms, trying to trace the possible linkage between being "good" and performing well. They found that the "evidence that socially responsible firms have lower discount rates, and thereby investors have lower expected returns, is stronger than the evidence that socially responsible firms deliver higher profits or growth."[4] This is the opposite of what most ESG proponents believe, of course. Cornell and Damodaran state, "There are firms that benefit from being socially responsible, but there are just as clearly firms where being socially responsible creates costs with no offsetting benefits." Their conclusion is significant: "Telling firms that being socially responsible will deliver higher growth, profits and value is false advertising." Lower dividend payments for favored companies and higher dividend payments for excluded ones is simple common sense.

Of course, none of these arguments is likely to stop Reclaim Finance and others who sympathize with them. They will continue to accost financial intermediaries who affiliate with any company they regard as insufficiently civic-minded. Still, we now appreciate the limited efficacy of this approach. Campaigns insisting upon divestiture draw impactful public attention to companies and industries. Companies that are the direct targets of their ire and the financial institutions that serve them are forced to take some notice. No one welcomes negative publicity. Adani has started pushing harder for renewable power and changed their financial practices, victories for which StopAdani's activists no doubt deserve some credit. But Reclaim Finance and their sympathizers will never achieve all of their financial objectives. Capital will undoubtedly continue to flow to businesses and practices they despise. In the process, they may well help enrich those they most hope to dissuade.

Whereas StopAdani attacks one corporation and those who enable it, and while Reclaim Finance villainizes every financial services firm that doesn't meet their impracticable demands, another prominent activist movement, Fridays for Future, has a very different target: global policy-makers. It also has a very different, animating demographic: schoolchildren. Started by the inestimable Greta Thunberg, Fridays for Future seeks "to put moral pressure on policy-makers, to make them listen to the scientists, and then take forceful action to limit global warming."

Fridays for Future's website is headlined with the hashtag #uprootthe-system. It enumerates three demands from their so-called Declaration of Lausanne: "(1) Keep the global temperature rise below 1.5°C compared to preindustrial levels; (2) ensure climate justice and equity; and (3) listen to the best united science currently available." Since August 2012, Fridays for Future has helped organize more than one hundred thousand youth-led strikes in 212 countries around the globe. At only four foot, eleven inches and nineteen years of age, Greta Thunberg has also secured three consecutive nominations for the Nobel Peace Prize, been awarded an honorary fellowship by the Royal Scottish Geographical Society, and been included in *Forbes*'s list of the "World's One Hundred Most Powerful Women." She also became *Time* magazine's youngest "Person of the Year." As improbable as it may be, there is no doubt Greta Thunberg has struck a deeply resonant chord (figure 3.2).

Policy-makers have taken note of Ms. Thunberg and her fellow young activists as well. After a ninety-minute meeting alone with then German chancellor, Angela Merkel, on August 20, 2020, the chancellor's office released the following statement: "Both sides agreed that global warming is a global challenge that industrialized countries have a special responsibility

FIGURE 3.2. Don't mess with Greta!
Greta Thunberg. NICHOLAS KAMM / Contributor, Getty Images.

to tackle. The basis for this is the consistent implementation of the Paris Climate Agreement."

Both sides: Chancellor Merkel on behalf of all industrialized nations and Greta's, that is. True to form, Thunberg's assessment of the meeting was curter: "We want leaders to step up, take responsibility, and treat the climate crisis like a crisis."

Greta Thunberg's success has helped spark many other adult-led activist organizations to lobby similarly for political change. Demand Progress, the Good Lobby (based in Europe), the Sunrise Movement, and Extinction Rebellion, among others, combine public protests with targeted marketing campaigns for or against public policies and politicians they support or reject. Some of these movements—like Black Lives Matter—are highly targeted, focusing on specific issues like racial injustice or climate concerns. Others, like Operation Libero in Switzerland, embrace multiple causes, ranging from marriage equality and migrant rights to climate action and higher corporate tax rates. Each seeks specific social and environmental changes. And each expects business and finance to support them and the attainment of their goals.

Many other organizations have similar aims but use different approaches. For example, rather than look to those in positions of influence to solve shared environmental and social problems, some movements and activist organizations advocate a more bottom-up approach. For these NGOs—organizations like the Green Belt Movement (GBM) in Kenya, the Climate Reality Project, and Charter for Compassion to name just a few—impacts on socioeconomic outcomes are sought through the mobilization of millions, if not billions, of individual actors and actions. If you want to fix macro problems, these bottom-up organizations contend, don't fight just to change the system. Strive simultaneously to change *individual* behaviors that ultimately drive the wheels of commerce, including your own. In short, be the change you want to see in the world, and encourage others to do the same.

As is often the case with impactful NGOs, GBM grew out of local, visionary, and inspirational female leadership. GBM was founded by the Nobel Prize–winning professor and author Wangari Maathai as an offshoot of her National Council of Women of Kenya. As the penultimate "frontline workers," Kenyan mothers bore painful witness to how climate change led their freshwater streams to dry up and their food sources to become less secure. Rather than accept environmental degradation as the unavoidable price to pay for economic development, GBM encouraged women to fight back *with their bare hands*. Specifically, GBM has championed a highly successful reforestation initiative that could ultimately span eight thousand kilometers across the entire width of Africa. To date, GBM volunteers have nurtured seedlings and planted more than fifty million trees. These trees have helped bind soil, capture carbon, and store rainwater. GBM activists have also popularized simple environmental practices, including using firewood more sparingly. More than four thousand community groups have become involved, scaling up proven methods that make a visible and direct difference in their local ecosystems. As a result, hundreds of river streams and more abundant crops have started to come back. In short, GBM's methods are succeeding. GBM is what I call an "exemplar of hope." GBM is a notable community-led example of how best to nurture more inclusive, sustainable growth in a highly scalable manner.

Similarly, after receiving the TED Prize in 2008, the best-selling author and acclaimed scholar Karen Armstrong voiced her wish to create and sustain a more viable and united human family. Armstrong cemented her efforts in a novel global "Charter for Compassion." Her charter encourages more and more people, artists, schools, businesses, cities, and ultimately large communities to live by the Golden Rule. The Golden Rule is based upon the inviolable principle of human dignity. It requires one to treat all with the same level of justice, equity, and respect one wishes for oneself. A corollary of this edict is to inflict no pain and to treat others and one's circumstances with unwavering compassion. A less apparent requirement of compassionate living is to coexist in complete harmony with the environment so as to not ruin it for future generations.[5] The environmental component of the Charter for Compassion focuses on clean water and sanitation, responsible consumption and production, meaningful climate action, and the preservation of life on land: four of the seventeen Sustainable Development Goals of the United Nations, which we will soon discuss.

While the activist initiatives I have enumerated strive to change the system fundamentally—either by targeting offending corporations, their financiers, overly compliant politicians, or underlying personal behaviors—there is a final type of activist effort worth highlighting, one that takes the opposite approach. Rather than trying to change the trajectories of corporate activities from the *outside*, this strategy seeks to ignite change from *within* corporate boardrooms. Activists seek to do so through the formal adoption of shareholder resolutions and the election of new board directors with reformist credentials. Such organizations, like Engine No. 1, buy the shares of companies they want to change rather than sell them, and they become insider activists. In their view, corporations are best redirected and renewed without violent protests and bombast. Rather, they seek strategic change through engaged share ownership.

Engine No. 1's home page describes the firm as "an investment firm purpose-built to create long-term value and bring common sense back to capitalism." In effect, their strategy is to save capitalism by being more engaged capitalists: "We believe a company's performance is greatly enhanced by the investments it makes in workers, communities, and the environment. Moreover, we believe that over time, the interests of Main Street and Wall Street align—and we can engage as active owners to create value by focusing on this alignment."

Though only a few years old, Engine No. 1 has already made an impact. After years of ignoring activist demands to diversify their core business strategy away from oil and gas to other forms of cleaner, more sustainable power, Exxon Mobil was forced to add three new directors to its board through a campaign Engine No. 1 designed and led. The purpose of their campaign, dubbed "Reenergize Exxon," was not to starve Exxon of capital—a strategy with limited efficacy, as we have discussed. Instead, it was to attempt to enhance value for all of Exxon's shareholders and the planet by getting Exxon's management to increase investments in clean energy, improve long-term capital allocation, and promote executive leadership with proven capacity to navigate the pending emissions transition. Where other activists failed, Engine No. 1 succeeded. That said, it might have been even more valuable had it happened sooner. Over the ten years leading up to Engine No. 1's campaign, Exxon's shareholder returns were negative 15 percent, including dividends. This weak performance stands in contrast to the approximate positive 300 percent return for the S&P 500 over the same period. In other words, Exxon's long-term strategy not only penalized their employees, suppliers, communities, and the environment; it damaged their shareholders as well.

As this discussion has shown, corporate America has been served notice by protestors and agitators of nearly every interest group, geography, and demographic. With growing numbers of violent, often devasting storms—social and environmental—many global CEOs have found themselves flat-footed and increasingly on the defensive. Senator James E. Watson—a Republican from Indiana who lost his seat like many others in 1932 to the

Democrat Frederick Van Nuys because of FDR's long coattails—is credited with originating the familiar phrase "If you can't beat 'em, join 'em."

As we turn our attention to an enlarging cadre of C-suite insurrectionists who have embraced many of the activist arguments described in this chapter, "join 'em" very much appears to be what many global business leaders have decided to do. Will verbal commitments from this growing chorus of CEOs result in more inclusive, sustainable prosperity—or will their deeds fail to match their words?

You be the judge.

Chapter Four

C-SUITE INSURRECTIONISTS

It's a question of whether society trusts you or not. We need
society to accept what it is that we do.
GINNI ROMETTY, CEO, IBM

Making money comes from solving problems. Solving problems
does not come from making money.
PROFESSOR COLIN MAYER, UNIVERSITY OF OXFORD

If America's business community has had a genuine "Road to Damascus"
moment, it appears to have taken place sometime before the summer of
2019.[1] It was then that the Business Roundtable (BRT)—a nonprofit association of American CEOs whose businesses employ twenty million people,
generate more than $10 trillion in revenues, and comprise $20 trillion in
stock market capitalization—revised and restated the purpose of their organizations. Their new, highly publicized credo deliberately adopted a much
broader commitment to serving each of their stakeholders.

Since its inception in 1972, the BRT has understood its principal challenge is to convince the broader public of business's inherent importance and
value. As cofounder John Harper, CEO of Alcoa, stated then, "Business must
take an active, aggressive role in developing understanding of and support
for the free-market system by reestablishing the public's confidence." Back
in 1972—just a few years into America's lost economic decade—businesses
were seen more as problem creators than problem solvers, much the same
criticism being made today. As they struggled to please more folks in the
decade that followed, their commercial acumen flagged. In its first decades,
the BRT operated like a self-regulatory body primarily devoted to improving corporate governance principles. The BRT emphasized the primary role
boards must play in delivering improved shareholder value in a series of
white papers and pronouncements. In 1997, the BRT issued a pointed statement about business's primary responsibilities being to shareholders (which

they then called stockholders); the interests of stakeholders were considered subsidiary. In line with our review of the stakeholder-versus-shareholder debate, in their 1997 edict, they claimed there was no conflict between their mutual claimants *over the longer run*:

> Some say corporations should be managed purely in the interests of stock-holders or, more precisely, in the interests of [their] present and future stock-holders over the long term. Others claim that directors should also take into account the interests of other "stakeholders" such as employees, customers, suppliers, creditors and the community. The Business Roundtable does not view these positions as being in conflict. . . .
>
> It is in the long-term interests of stockholders for a corporation to treat its employees well, to serve its customers well, to encourage its suppliers to continue to supply it, to honor its debts, and to have a reputation for civic responsibility. . . .
>
> In the Business Roundtable's view, the paramount duty of management and of boards of directors is to the corporation's stockholders; the interests of other stakeholders are only relevant as a derivative of this duty. The notion that the board must somehow balance the interests of stockholders against the interests of other stakeholders fundamentally misconstrues the role of directors. It is, moreover, an unworkable notion because it would leave the board with no criterion for resolving conflicts between the interests of stockholders and of other stakeholders or among different groups of stakeholders.

Clearly, the 1997 position of the BRT, speaking on behalf of corporate America, was that stakeholders do indeed matter but only insofar as their interests serve the long-run interests of shareholders. If the former are ever in conflict with the latter, it is the corporate boards' responsibility to ensure they are adjudicated in favor of shareholders *over the long run*. Attempting the mollification of other stakeholders over shareholders in real time was not just hard; according to the BRT, it was "unworkable." In 1997, the BRT claimed corporate boards had no such Solomonic wisdom or superpower to weigh one stakeholder's interest against another's. For them, there was one clear goal: *shareholders' long-term interests*.

Fast-forward twenty-two years and the BRT felt compelled to restate and diverge from their essentially Friedmanesque position of 1997 in several important ways. In their 2019 restatement on the purpose of a corporation, BRT's chairman, Jamie Dimon—then also the chairman and CEO of JPMorgan Chase—and BRT's Corporate Governance Committee chair, Alex Gorsky—then the chairman and CEO of Johnson & Johnson—went to exceptional lengths to describe how different BRT's new principles were. "This new statement better reflects the way corporations can and should operate today," Gorsky claimed. "It affirms the essential role corporations can play in improving our society when CEOs are truly committed to meeting the needs of all stakeholders." "Our revised statement is an acknowledgment that business can do more to help the average American," Jamie Dimon added. But exactly how different is the BRT's 2019 position on corporate purpose versus the vision they laid out in 1997?

BRT'S 2019 STATEMENT ON THE PURPOSE OF A CORPORATION

We believe the free-market system is the best means of generating good jobs, a strong and sustainable economy, innovation, a healthy environment, and economic opportunity for all. . . .

While each of our individual companies serves its own corporate purpose, we share a fundamental commitment to all of our stakeholders. We commit to:

- Delivering value to our customers. . . .
- Investing in our employees. . . .
- Dealing fairly and ethically with our suppliers. . . .
- Supporting the communities in which we work. . . .
- Generating long-term value for shareholders. . . .

Each of our stakeholders is essential. We commit to deliver value to all of them, for the future success of our companies, our communities, and our country.

With their additional emphasis on employees, suppliers, and the communities in which they operate, the BRT seems to have raised their 1997 bar in terms of responsibility and scope. "The traditional role of the BRT has been policies that promote economic growth," the president of BRT, Joshua Bolten, explained. "We haven't lost that. But we've added an 'opportunity agenda,' recognizing the need for policies that better spread the benefits of economic growth." In other words, Bolten is saying, BRT believes economic growth must be more inclusive. They also nod toward the importance of climate issues, explicitly mentioning the importance of protecting the environment and embracing sustainability. They made no explicit commitments to carbon reduction, however, let alone alignment with the Paris Agreement. Refusing to weigh in more precisely on the climate debate suggests significant differences remain among their members on this topic. There were no apparent, disqualifying disagreements on customers, employees, suppliers, local communities, or more long-term value for shareholders, however.

"I welcome this thoughtful statement by Business Roundtable CEOs on the purpose of a corporation," Bill McNabb, then the CEO of Vanguard, observed upon launch. "By taking a broader, more complete view of corporate purpose, boards can focus on creating long-term value, better serving everyone—investors, employees, communities, suppliers, and customers."

Most print and broadcast media were even more effusive in their praise than McNabb. The *Wall Street Journal* described the BRT's new position on corporate purpose as "a major philosophical shift," while the *Washington Post* said it was "the loudest reform call yet from inside the system, . . . a potential sea change . . . so significant and so welcome." The *New York Times* said it represented "a significant shift that breaks from decades of long-held corporate orthodoxy." *USA Today* and *NBC News* went the furthest: it was "a stunning new mission statement" and "something seismic."

Others were less convinced.

"What I see are well-meaning activities that are little more than virtuous side hustles," the social commentator Anand Giridharadas told the *Fortune* magazine CEO, Alan Murray, in an article on the BRT's new doctrine. "Many companies are focused on doing more good but less attentive to doing less harm."[2] Giridharadas's best-selling book *Winners Take All: The Elite Charade of Changing the World* drew attention for its compelling argument that societal problems would best be solved by public policy rather than part-time, compromised efforts by businesses. As the ultimate objective of businesses is to generate profits, Giridharadas believes businesses are both unfit and unlikely to generate comprehensive social and environmental change: "When a society helps people through its shared democratic institutions, it does so on behalf of all, and in a context of equality. When a society solves a problem politically and systemically, it is speaking on behalf of every citizen." The progressive standard-bearer Senator Elizabeth Warren harbors beliefs similar to Giridharadas's but states them more actionably: "If Jamie Dimon thinks it's a good idea for giant corporations to have multiple obligations, he and I agree. But then, let's make them law."

Warren and Giridharadas may be surprised, but their skepticism over the genuine conversion of business leaders to bona fide "stakeholder enthusiasts" is shared by several notable academics, including a number of reliably conservative ones. "The statement is largely a rhetorical public relations move, rather than the harbinger of meaningful change," wrote Roberto Tallarita and Lucian Bebchuk of Harvard Law School. "The support of corporate leaders and their advisers for stakeholders is motivated by a desire to obtain insulation from hedge fund activists and institutional investors. But stakeholderism does not benefit stakeholders, shareholders, or society. If stakeholder interests are to be taken seriously, stakeholderism should be rejected."[3] Bebchuk and Tallarita believe businesses should remain immutably focused on the bottom line. As the BRT asserted in 1997, businesses should not attempt the impossible by requiring board members to make repeated tough trade-off calls between competing stakeholders with highly uncertain outcomes. Like Giridharadas, Bebchuk and Tallarita also feel public policy, regulation, and individual behavior are better suited for social engineering, not an ever-morphing list of presumed corporate responsibilities. Corporations can't and won't balance competing stakeholder interests optimally: it's just too hard to do. Bebchuk and Tallarita are also not alone. The corporate consultant Andrew Winston—a coauthor of the best-selling book *Green to Gold*—also views the new BRT statement derisively: "This new discussion of purpose is good, and it mirrors what some big investors are saying. But we need a much bigger pivot to circular, renewable-energy-based business models that value the long term, protect natural capital, and invest in human development and equality. That level of change is currently light-years beyond the BRT statement."[4]

By claiming the BRT's statement mirrors what some big investors are saying, Winston appears to be referring at least in part to BlackRock chairman and CEO, Larry Fink. He would be right to do so. The BRT was neither the first nor even the loudest C-suite advocate for broader corporate responsibility in the past decade. By most measures, including casual Google searches, that role appears to belong to Fink. Larry Fink was a signatory of the BRT's 2019 restatement of corporate purpose on behalf of the thirty-one-year-old firm he founded and leads. Arguably, he also has been the most vocal CEO globally about governmental failures to address growing social challenges. Fink has repeatedly stressed the need for finance and business to help close these gaps. And the primary mechanism Fink has used to exhort more impactful corporate behavior has been his widely quoted annual letter to CEOs.

It's worth spending a few minutes on Larry Fink's annual CEO letters. Examined closely, they reflect the arc of the stakeholder capitalism debate over the last decade. They have been distributed to thousands of corporate leaders around the globe every year since 2012. In them, Fink attempts to speak on behalf of the clients who have asked BlackRock to maximize their returns not over the next quarter or two but rather over decades. A majority of the $10 trillion or so BlackRock manages is on behalf of individual retirees, insurance policyholders, public pension plans, endowments, and foundations—that is, institutions with long-term liabilities.

Fink writes his CEO letters first and foremost for these individuals—not for himself, and certainly not for his personal values. His initial CEO letter was only 260 words long. It focused on the importance of governance and BlackRock's desire to engage more directly with corporate leadership on their long-term plans and responsibilities. His 2021 letter was more than two thousand words long, not including detailed footnotes. Clearly, Fink has found more things to talk about over the years.

Before dissecting his annual messages, it is also helpful to say a few words about BlackRock's clients and why Larry Fink and his colleagues obsess over their financial well-being.

The pensioners who rely upon BlackRock to provide for their retirement include billions of folks from every walk of life: coal mine workers and airline pilots from red, blue, and purple states; central bankers and devotees of every Christian, Muslim, and Jewish tradition imaginable; public school teachers and American military servicemen and -women from thousands of communities; citizens of more than 110 countries; millions of U.S. state and federal public employees; hundreds of thousands of employees of oil, gas, and mineral extraction companies; frontline health care workers and industrial union members; and countless retired citizens who depend upon their funded federal pension plans in Abu Dhabi, Australia, Canada, Chile, China, France, Japan, Kazakhstan, New Zealand, Norway, Saudi Arabia, Singapore, South Africa, Sweden, and Taiwan, to name only those for which public data is available. Equally important, thousands of companies rely on BlackRock's portfolio managers and stewards to buy their debt instruments and vote on board resolutions. Moreover, hundreds of finance ministers, political leaders, multilateral organization officials, and central bankers value BlackRock's macroeconomic counsel on how they might best achieve inclusive, sustainable growth.

Much is made today of the existential challenge of climate risk and the urgent need to promote greater economic mobility and social justice, all for good reason. But an equally dire and potentially catastrophic problem concerns our looming retirement crisis. As the economist George Magnus has memorably said, humans now live in the "age of aging."[5] In sum, we are living longer, having fewer babies, and—in countries like Japan and in much of Europe—not allowing enough immigration to replace those souls that are being lost. The net effect in many countries is ever older, shrinking populations and more burdened workforces. Soon these debilitating demographic challenges will descend over much of the globe— including China and, if we are not farsighted enough, the United States.[6] Unfortunately, with the possible exception of Norway, the Netherlands, and perhaps a Gulf nation or two, no country has saved enough money to support their growing numbers of retirees in comfort in perpetuity.

While governments are financially unready for their retirees, even fewer individuals have supplemented their savings adequately to make up for unavoidable, impending social security shortages. One recent study finds half of all American households will not have enough retirement income to maintain their preretirement standard of living. Scrimping will be necessary even if most work to the age of sixty-five and annuitize all their financial assets, including by taking reverse mortgages on their homes.[7] Meanwhile, a plurality of older Americans will have nothing more than Social Security benefits to live on; for them, all they can hope for in their retirement is subsistence through federal support, family assistance, and occasional charity. In fact, fewer than 10 percent of the globe's inhabitants can expect to maintain their standard of living after retirement. Neither governments, nor employers, nor individuals have adequately saved to care for the legions of aged and infirm soon to retire. Of the nine billion people expected to be alive by 2050, when the Paris goals are supposed to be reached, about one-quarter will be over the age of sixty-five. As figure 4.1 suggests, unless the retirement assets that BlackRock and all other asset managers oversee manage to return something like 7 to 8 percent a year over the coming decades, few countries will even be able to provide food, housing, and decent health care to their older citizens. Countries with aging demographics and so-called pay-as-you-go retirement plans will be especially challenged in upholding their promises to their citizens for basic daily needs and medical care. An open question for the decades immediately ahead is what percentage of our parents and grandparents will genuinely be able to retire with dignity.[8] It is a fair assumption that millions will suffer unnecessarily if we do not plan ahead and take the right steps today.

System	Overall Index Value	Sub-Index Values		
		Adequacy	Sustainability	Integrity
Iceland	84.2	82.7	84.6	86.0
Netherlands	83.5	82.3	81.6	87.9
Denmark	82.0	81.1	83.5	81.4
Israel	77.1	73.6	76.1	83.9
Norway	75.2	81.2	57.4	90.2
Australia	75.0	67.4	75.7	86.3
Finland	73.3	71.4	61.5	93.1
Sweden	72.9	67.8	73.7	80.0
UK	71.6	73.9	59.8	84.4
Singapore	70.7	73.5	59.8	81.5
Switzerland	70.0	65.4	67.2	81.3
Canada	69.8	69.0	65.7	76.7
Ireland	68.3	78.0	47.4	82.1
Germany	67.9	79.3	45.4	81.2
New Zealand	67.4	61.8	62.5	83.2
Chile	67.0	57.6	68.8	79.3
Belgium	64.5	74.9	36.3	87.4
Hong Kong SAR	61.8	55.1	51.1	87.7
USA	61.4	60.9	63.6	59.2
Uruguay	60.7	62.1	49.2	74.4
France	60.5	79.1	41.8	56.8
Malaysia	59.6	50.6	57.5	76.8
UAE	59.6	59.7	50.2	72.6
Spain	58.6	72.9	28.1	78.3
Colombia	58.4	62.0	46.2	69.8
Saudi Arabia	58.1	61.7	50.9	62.5
Poland	55.2	60.9	41.3	65.6
China	55.1	62.6	43.5	59.4
Peru	55.0	58.8	44.2	64.1
Brazil	54.7	71.2	24.1	71.2
South Africa	53.6	44.3	46.5	78.5
Italy	53.4	68.2	21.3	74.9
Austria	53.0	65.3	23.5	74.5
Taiwan	51.8	40.8	51.9	69.3
Indonesia	50.4	44.7	43.6	69.2
Japan	49.8	52.9	37.5	61.9
Mexico	49.0	47.3	54.7	43.8
Korea	48.3	43.4	52.7	50.0
Turkey	45.8	47.7	28.6	66.7
India	43.3	33.5	41.8	61.0
Philippines	42.7	38.9	52.5	35.0
Argentina	41.5	52.7	27.7	43.0
Thailand	40.6	35.2	40.0	50.0
Average	61.0	62.2	51.7	72.1

FIGURE 4.1. The age of aging: no country is 100 percent ready for its retirees.

Reprinted with permission from Mercer (2021), Mercer CFA Institute Global Pension Index. Available at: www.mercer.com/globalpensionindex.

Surveyed closely, the topics of Larry Fink's CEO letters have always reflected these grave human concerns and the shared social urgency of generating superior long-term investment outcomes. If the global economy does not continue to grow, or if pension savings do not continue to increase in value relative to inflation, it really won't matter so much what the global temperature is: millions, if not billions, will be unable to feed, clothe, and/ or shelter themselves. Fink first started writing his letters to address a widespread problem laid bare by the 2008 global financial crisis: corporations were being insufficiently transparent about the risks they were carrying on their balance sheets. Over the past decade, many of the troubles that corporations like General Electric have had to contend with included a tangled web of financial exposures mindlessly retained on their balance sheets. GE's was a significant failure for a company that supposedly excelled at aircraft parts, power generation, and health care. Enron, WorldCom, and other companies similarly failed because they strayed from their core business expertise and made disastrous forays into finance, where presumed riches proved illusory.

As the memory of the global financial crisis faded, it became clear that too few companies were planning for and investing in their future growth. Even fewer were thinking about their multiple stakeholders. Viewed from his global perch, with a multinational lens few others share, Fink could see how some corporations were better at realigning their strategies to capture emerging consumer preferences and needs than others. He could also see how some countries were more farsighted in terms of public investment in education, infrastructure, and strategic industries than others. In response, improved governance and greater attention to strategic long-term growth plans have been the most prominent opportunities Fink's letters have attempted to highlight. In early 2014, he wrote,

> Dividends and buybacks in the U.S. alone totaled more than $900 billion, the highest on record. With interest rates approaching zero, returning excessive amounts of capital to investors sends a discouraging message about a company's ability to use its resources wisely and develop a coherent plan to create value over the long term.

Excoriating companies for sending back too much cash to their shareholders may seem like a counterintuitive way to build an asset management business. But Fink's criticism is understandable given his primary objective has been to maximize client returns over decades rather than a few months or quarters. When one's goals span generations, one's preoccupations and concerns acquire a different focus and tenor. Looking at the multiple challenges humanity will face in the decades to come, it seems supporting retirees should be a top priority—ostensibly above the environmental challenges we simultaneously face. Why? Because underfunded retirements are a certainty, whereas continued climate woes and their human costs are merely high probabilities.

As figure 4.2 shows, no Fink-to-CEO letter struck a more resonant chord than his 2018 message on corporate *purpose*. That letter included several assertions and significant urgings:

To prosper over time, every company must deliver financial performance and show how it positively contributes to society.

Without a sense of purpose, no company, either public or private, can achieve its full potential. Companies without purpose will provide subpar returns to the investors who depend on them to finance their retirement, home purchases, or higher education.

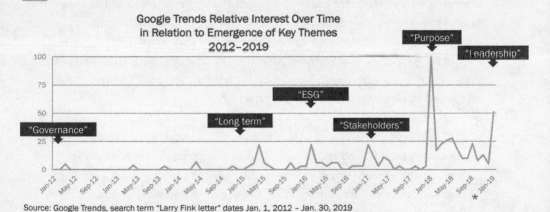

FIGURE 4.2. Measures of interest in Larry Fink's CEO letters.
Courtesy of Globescan.

All companies must ask themselves: What role do we play in the community? How are we managing our impact on the environment? Are we working to create a diverse workforce? Are we adapting to technological change? Are we providing the retraining and opportunities that our employees and our business will need to adjust to an increasingly automated world? Are we using behavioral finance and other tools to prepare workers for retirement, so that they invest in a way that will help them achieve their goals?

It is hard to overstate the resonance of this forceful "corporate purpose theme." Injecting legitimate purpose into a corporation's ethos not only raises the stakes, but it also often generates dividends for all. "Purpose solves the problem for shareholders as well as stakeholders," Professor Colin Mayer of the University of Oxford has observed. "It says what a company is not there to do as well as what it is there to do. It specifies in whose interests the company is being run and what its values are. It simplifies and makes precise what the 'only shareholders matter' philosophy leaves vague and confused."

One example of Mayer's sweeping claim can be found in the recent actions of the 112-year-old General Motors. In 2017, General Motors condensed its social mission statement into six concise words: "Zero crashes, zero emissions, zero congestion." Tight, prosaic, compelling. GM's leadership further backed up these words with action. GM's new Ultium platform pledged to put an electric vehicle within every consumer's reach, just like Henry Ford had promised with his Tin Lizzie a century earlier. "There was a huge outpouring of support," the GM CEO, Mary Barra, reported within weeks of launching the new campaign. It turns out GM's employees were much more motivated by making human lives better and the environment more sustainable than they could ever have been by striving to increase the company's next quarter's earnings by two cents per share. It also turns out clients want to do business with firms that share their values and aspirations by looking after their employees' well-being. *Why has it taken so long for corporations to understand this?* Every human being seeks meaning and fulfillment. This includes every employee of every corporation. Businesses have an opportunity to tap into this yearning. When they do, they often multiply employee engagement and customer loyalty. Many of these benefits may accrue without spending an incremental dime, moreover. Corporations that develop and execute a real, life-enhancing purpose often transform themselves through vision and willpower. Purpose clarifies and motivates. And purposeful businesses often perform better than those without purpose, as Professor Mayer observes.

In Larry Fink, stakeholder capitalists appear to have found a CEO of CEOs as a standard-bearer. Stakeholder capitalists fundamentally believe short-term thinking has been the bane of what modern capitalism has been getting wrong, from the environment to fairer wages and working conditions to greater inclusion and diversity. Encouraging more long-term thinking through forcefully illustrating how such issues impact corporate performance and society at large seems all but certain to generate more inclusive, sustainable growth. Having a CEO who is a vocal stakeholder capitalist has also proven to be rewarding for BlackRock's shareholders. From 2005 through early 2022, BlackRock's stock had risen more than tenfold, a period during which the DJIA had merely tripled. Outperforming the broader market by more than three times is a stunning achievement. Fighting short-termism is also at the heart of all ESG investing, implicitly if not explicitly. If more corporations were genuinely focused on the longer term, environmental, social, and governance concerns would be more top of mind. And if ESG thinking becomes second nature, more commonsense long-term solutions may be conceived and implemented. An ESG ethos should help foster inclusive, sustainable growth.

But like stakeholder capitalism itself, Larry Fink's embrace of all things long term and sustainable has not been linear. Sometime during 2019, Fink appears to have had an additional epiphany: identifying the environment and climate as merely one of several topics of concern wasn't dispositive enough. As he listened more closely to what clients were telling him, Fink concluded that he and his colleagues had to adopt a more powerful position on climate. Fink's 2020 letter to CEOs displayed this newfound conviction:

The evidence on climate risk is compelling investors to reassess core assumptions about modern finance. Research . . . is deepening our understanding of how climate risk will impact both our physical world and the global system that finances economic growth. Will cities, for example, be able to afford their infrastructure needs as climate risk reshapes the market for municipal bonds? What will happen to the thirty-year mortgage—a key building block of finance—if lenders can't estimate the impact of climate risk over such a long timeline, and if there is no viable market for flood or fire insurance in impacted areas? What happens to inflation, and in turn interest rates, if the cost of food climbs from drought and flooding? How can we model economic growth if emerging markets see their productivity decline due to extreme heat and other climate impacts?

Climate risk is investment risk. Every government, company, and shareholder must confront climate change.

Though unapparent to the outside world at the time, these excerpts from Larry Fink's 2020 letter marked one of the more significant and strategic pivots in his firm's thirty-year history, equal perhaps to launching its risk analytics platform Aladdin in 1999, or its bold purchase of iShares from Barclays in 2009. 2020 was the first year Fink announced to his colleagues and to the world that all investment risk models should systematically account for climate change. Investment analytics that did not incorporate climate as a specific stand-alone risk were flawed and inadequate tools for allocating capital. As a result, sustainable investing—that is to say, investment decision-making based upon a deliberate effort to understand how environmental, social, and governance concerns may impact asset prices—would henceforth become a new encompassing standard for all product development and active portfolio management at BlackRock. This new risk paradigm would enhance some prior practices while leapfrogging others.

The impact of Fink's decision to single out climate risk was seismic. Other asset management firms had taken stronger stances on climate risk earlier than BlackRock had—but added together, those firms represented a fraction of BlackRock's multi-trillion-dollar size. This shift reverberated through every corporate boardroom, investment desk, regulatory agency, and legislative body in the world. It also reverberated through BlackRock. Default risk had long been factored into all of BlackRock's credit models. Similarly, prepayment risks have always been accounted for in every mortgage-backed security exposure. Forevermore, ESG risks would be essential inputs for all of the firm's active capital commitments. No asset class—public or private, debt, equity, or cash—would escape this additional scrutiny. No portfolio manager could be a climate change denier or contrarian on climate, either. The risks of environmental, social, and governance concerns would be incorporated into every BlackRock portfolio manager's investment decisions because they represent real risks.

But Larry Fink's 2020 letter actually did more than this. It simultaneously predicted that the financial impact of climate risk would be felt much sooner than outdoor thermometers could possibly suggest. One further impact is that investors would start reallocating capital in anticipation of changes that seem increasingly likely, impacting some climate-sensitive asset prices much sooner, if not immediately. And over time, more changes would be driven by government policy decisions, like mandating more disclosures, putting a use tax on carbon, and relabeling investment products. Others would be deeply impacted by shifting consumer and investor preferences. Markets are forward-looking and surprisingly efficient, after all; they must discount the growing likelihood of climate risks before climate change fully manifests. As the soaring share prices of Tesla and Beyond Meat revealed soon after, it was not the first time that one of Larry Fink's hunches proved prescient.

And, as you might expect of any demanding CEO who has built and led a highly successful company, the standards Fink advocated for other companies were applied to the home team first. BlackRock's accounting team was required to implement all relevant climate-related financial disclosures and Sustainable Accounting Standards Board (SASB) guidelines before other companies were encouraged to do the same. Charges of hypocrisy—"do as I say, not as I do"—are an all-too-common fate for many folks who make a living in part by suggesting how others should behave. In 2019, BlackRock began reporting its results in line with the Task Force on Climate-Related Financial Disclosures (TCFD) and the SASB. In both 2020 and 2021, Black-Rock's annual report also disclosed a carbon-neutral corporate footprint, as the firm had purchased carbon credits to offset emissions associated with its operations.

That said, no communication from BlackRock raised the bar higher or challenged all of Fink's colleagues and fellow Group Executive Committee members more than his 2021 letter committing all of the firm's analytics and investments to the goal of net-zero greenhouse gas emissions by 2050. The concept of net zero is complicated, and its prospects for successful implementation remain highly uncertain. Net zero effectively requires corporations and societies put no more carbon into the atmosphere than they remove by other means over a specific time horizon, usually by 2050. On its face, daunting challenges seem to suggest this may be impossible. How can corporations accurately measure how much carbon they are putting into and taking out of the atmosphere? Should they measure only their direct emissions (scope 1) or include indirect emissions and those of their suppliers (scopes 2 and 3, respectively)? If they do not know what these are, how much time should they be given to figure it out? No firm can promise to manage what it and no one else can yet measure. To this net-zero pledge, BlackRock soon added two others: a new principle of "heightened scrutiny" involving all of their investment stewardship practices and the further promise of "temperature alignment metrics" on all of their funds, wherever qualitative and quantitative data allow. "Heightened scrutiny" means staying on top of corporations that are not doing enough to safeguard the environment, their workers, and their long-run growth prospects. Thus, if a public corporation is making an insufficient effort to navigate the goals and demands of a net-zero economy, mistreating their workers, or allocating their capital inefficiently, they may well hear from BlackRock's stewardship team. If they remain intransigent after those conversations, they may have board members challenged or suffer defeats in future proxy fights. Fink's 2021 letter to CEOs left no doubt about what was expected:

> Given how central the energy transition will be to every company's growth prospects, we are asking companies to disclose a plan for how their business model will be compatible with a net zero economy—that is, one where global warming is limited to well below 2°C, consistent with a global aspiration of net zero greenhouse gas emissions by 2050. We are asking you to disclose how this plan is incorporated into your long-term strategy and reviewed by your board of directors.

While heightened scrutiny is more of an external initiative—one focused on the companies BlackRock invests in on behalf of its clients—the effort to assign a temperature alignment metric to each product and strategy is strictly internal at BlackRock. It refers to the percentage of companies in an index or fund that are verifiably aligned with achieving the goals of the Paris Agreement. Given the absence of reliable data, this is an exceptionally ambitious goal. Indeed, because of the lack of reliable data on corporate emissions, it is hard to know when or even if this objective will be achieved at scale. If corporations don't report the data needed, reporting alignment as promised will be very hard, if not impossible. No matter. By making this bold promise, Fink made sure extreme pressure to deliver would be felt inside BlackRock, no less than outside.[9]

Of course, BlackRock is not the only asset management firm keen to anchor its investment practices to more complete and rigorous climate-related data. Most, if not all, of the seven thousand signatories to the UN's Principles for Responsible Investment would like to do the same. Their success depends in no small part on the willingness and ability of corporations to adopt the aforementioned SASB and TCFD standards, which we will later scrutinize. Given these aspirations, it is also no surprise direct engagement by many corporate stewardship teams has been expanding exponentially. For example, BlackRock's stewardship team conducted more than 3,600 engagements with more than 2,300 public companies in fifty-five markets in the 2020/21 proxy voting year, a 50 percent increase versus the 2019/20 season. Their reviews covered more than 70 percent of the assets BlackRock manages on behalf of its clients. Explicit environmental engagements have multiplied in recent years, while social engagements—that is to say, corporate reviews on the "S" of "ESG"—have simultaneously doubled. Hundreds of carbon-intensive companies representing 60 percent of global scope 1 and 2 emissions[10] were actively reviewed face-to-face. Many faced adverse voting action in 2021 because of insufficient progress on climate and/or natural capital concerns. Corporations increasingly need to be the change their end investors want to see in the global economy—more inclusive, more sustainable, and, above all else, maniacal about plans for long-term growth. If not, growing numbers of investment stewardship teams will serve notice. The stakes are too high for anything less.

Among other priorities, many asset management firms are helping promote the net-zero target by getting more and more corporations on board. They are doing so not because of moral concerns, however. Many investment advisers are fighting to promote net-zero goals because of the deep conviction they share about optimizing long-term financial outcomes. Like many activists, a growing number of financiers and corporate CEOs believe humanity is in a race to save itself from an unknown and unknowable climate disaster. They are committing significant resources to do all they reasonably can to ensure the race against global warming is won. They are committing extensive resources because they believe retirees all around the globe need to avoid a climate disaster if they are to have any hope of retiring as they hope and deserve—in comfort, with dignity.

As mentioned earlier, like all matters relating to the global climate challenge, the concept of net zero is easy to explain but hard to implement. Hundreds of thousands of agents, activists, and investors are now calling upon every country and company to design and implement a plan showing exactly how they will emit no more carbon dioxide into the atmosphere than they remove by 2050, if not sooner. But note the word *net*. Pledges do not require zero use of carbon fuels; instead, *net* requires verifiable offsets to whatever carbon may be emitted. Net human-produced emissions need to decline by a stunning 8 to 10 percent annually for the next three decades starting *now* to have any chance of achieving Paris's goal of limiting the increase in global average temperature to 1.5 degrees Celsius above preindustrial levels. As my climate expert intoned in the introduction, there is no time to waste. Success depends critically upon what happens in the next few years. The 8 to 10 percent per annum pace of reduction may seem impossible, but COVID-19-induced restrictions on economic activity throughout 2020 slightly exceeded this amount. This gives a sense of how dramatically the global economy needs to be reordered just to be consistent with Paris.

No one should expect perpetual self-induced slowdowns to become the norm, however. In an ideal world, getting to net zero involves unprecedented efforts to reconfigure supply chains, modernize office space and equipment, telecommute more, travel less, create a more circular economy, power industry more sustainably, and much more. Getting to net zero will also require technological breakthroughs for clean energy production and consumption, including removing carbon from the atmosphere. The behaviors of every human being must change, from the energy they rely upon to the products they consume. No aspect of human activity will remain unaffected.

"Adam Smith was very clear in *The Wealth of Nations*," McKinsey's then senior partner Kevin Sneader said during a 2021 Davos panel discussion on how to lead a corporation in the twenty-first century. "The responsibility of a businessperson is to give to the community and enrich everyone." Sneader went on to cite widening income inequality, pressures from globalization, and the loss of public trust as three of many factors leading C-suite executives to join other activists for change: "You put all these together, [and] you get a cocktail that says, 'Time's up. We have to start doing things differently.'"

Sneader is another member of the corporate elite insisting it is time to change. But will his, BRT's, Fink's, and all other C-suites' efforts, combined with the aspirations of ESG investing proponents, change corporate behaviors sufficiently and quickly enough to make the goals of the Paris Agreement and other social goals attainable? This book is committed to identifying the proper role of business and finance to forge more inclusive, sustainable growth. But isn't it already too late to stop the worst effects of climate change from ravaging our ways of life, as well as our retirement portfolios?

Perhaps. This is why a more complete understanding of climate risk must be laid out in clear, unemotional, and scientific detail.

Chapter Five

WHAT IF +1°C = −$100 TRILLION?

The reality is climate change is real.
SYLVESTER TURNER, MAYOR OF HOUSTON

The warning from Tyler County Judge Jacques Blanchette on August 29, 2017, as Hurricane Harvey approached Houston was not one you could easily ignore: "Anyone who chooses not to heed this evacuation directive cannot expect to be rescued and should write their Social Security number in permanent marker on their arm so their bodies can be identified." Hurricane Harvey went on to register as the wettest hurricane in U.S. history, depositing sixty inches of rainfall over four days, an amount equivalent to the flow of Niagara Falls for four months straight. Sixty-eight people were killed, including the mother of a three-year-old girl who was found clinging to her mother's lifeless body. On top of the incalculable human carnage, more than three hundred thousand buildings and a half-million cars were damaged.

Climate change isn't *going* to happen. It is *already* happening. Just ask Mayor Turner.

Bill Gates's best-selling book *How to Avoid a Climate Disaster* helpfully focuses its readers' attention on the daunting challenge to redress climate change by highlighting two numbers: fifty-one billion and zero. The former represents the net amount of carbon measured in tons we humans collectively spew into the atmosphere *every year* (figure 5.1). The latter is where we need to get our net carbon emissions down to by 2050 to have much hope for limiting the increase in the global average temperature to well below two degrees Celsius versus the industrial era—the stated objective of

CARBON DIOXIDE OVER 800,000 YEARS

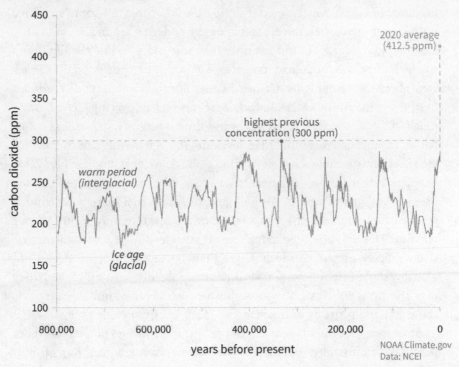

FIGURE 5.1. The earth's CO$_2$ levels are at historic highs.
NOAA.

the Paris Agreement. Gates's book is a sobering read. Given we have already gone up by about one degree Celsius, it appears we may have only one more degree to go.

One degree Celsius does not seem like very much. One could be forgiven for wondering, What's all the fuss?

It turns out that the only planet that astronomers and astrophysicists believe supports complex life within a radius of some one thousand light-years—ours, planet Earth—has had a total temperature gradation, from its lowest to its highest levels, of only ten degrees Celsius. That means that one to two degrees Celsius is 10 to 20 percent of the earth's real temperature change since life emerged.

That's obviously a lot, but it still does not put one to two degrees Celsius into full context. Honestly, how much difference might another one degree make?

During the last ice age, the earth was about six degrees Celsius colder than it is today. At that time, 30 percent of the earth was covered in ice, including most of North America. And during the Mesozoic era some sixty-five million years ago, when the dinosaurs were wandering around, it was a little less than four degrees Celsius warmer than it is today. While the dinosaurs roamed, there was, of course, no ice in Antarctica or Greenland. Sea levels were about sixty feet higher. Camels roamed around Canada, and crocodiles lived above the Arctic Circle. In fact, there was no ice covering Greenland or much of the Arctic and Antarctic the last time it was as warm as it is *today*. No ice then means that more than seven million square miles of existing ice are effectively unstable *now*—even if the globe doesn't get any warmer. What happens to that ice over the next ten, twenty, fifty, or one hundred years is pretty much anyone's guess. All we know for sure is that there will be much less of it unless the earth's temperature drops meaningfully soon. All our remaining ice may continue to melt slowly and steadily, or it may calve massively, quickly, even disastrously. Given there is practically no chance to end global warming right now, sea levels are undoubtedly heading higher. They may rise by as little as a few inches in the next century, but some models show they could rise by as much as forty feet. We simply don't know.

More than four billion people now live within fifty miles of the sea—more than half of humanity. Should sea levels rise even a few feet, lots of folks will need to move. In fact, in the absence of significant mitigation efforts, Bangkok, Calcutta, Dhaka, Guangzhou, Houston, Lagos, Miami, Mumbai, New Orleans, and Shanghai are just some cities that could become swamped. Today, these cities and their environs support trillions of dollars of public infrastructure and private property, as well as hundreds of millions of residents. More than a few of these cities could become unlivable. Some could even vanish (figure 5.2).

Sobering, no? As you dig further into what is required to bring those fifty-one billion tons of carbon down to zero, however, sobriety doesn't seem all that appealing; a stiff drink sounds much better. You know those multi-million-dollar oceanfront properties in the Hamptons and Miami and Bondi Beach? Well, they don't appear to be great long-term investments. In fact, if you are looking for beachfront properties, any listing sixty feet above sea level may well suffice. Be sure to take Denmark, Gambia, the Maldives, the Netherlands, and Qatar off your shopping list, though. These are the five lowest countries on Earth. Logistics in and out of their airports could prove highly challenging.

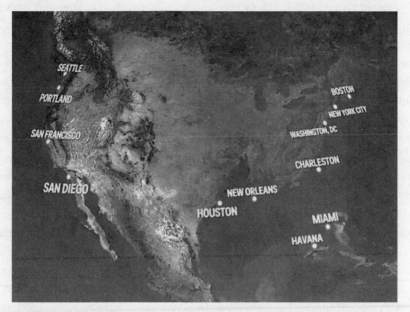

FIGURE 5.2. What might be left of the United States if *all* the ice melts.
Courtesy of Business Insider.

According to the data Bill Gates cites, if every motor vehicle on the planet—car and truck—were electrified tomorrow and if that electricity were sustainably generated—we would have effectively neutralized about 7 percent of those fifty-one billion tons. In other words, not very much. And if every home and commercial building on Earth were heated and cooled by, say, some combination of the sun, wind, and geothermal sources, also all renewable, we would have remediated another 8 percent. These would be astonishing achievements, of course, and these innovations have captured the imagination of every investor and policy-maker alike. But even if we did manage to achieve these two stunning feats—100 percent sustainably generated electrification of vehicles and 100 percent renewably sourced heating and cooling of every residence and commercial establishment—we'd still have another 43,350,000,000 tons of carbon per year to redress (figure 5.3). Electric vehicles and zero-carbon homes barely move the Paris needle. Much more comprehensive solutions are required if net zero is to be anything other than some bureaucrats' pipe dream.

Project Drawdown is one of several scientific efforts charting viable paths and policies to achieve net zero by 2050. They categorize their proposed solutions into four "waves": (1) quick wins, (2) new infrastructure, (3) growing natural sinks, and (4) deploying new tech. Among the quickest wins is ending all deforestation. Maintaining or extending existing flora appears compelling and seemingly easy to do. Rain forests help reverse the damage all our carbon consumption is causing. The only problem is that saving rain forests is not terribly popular where they are located, including Brazil and Indonesia. Without adequate compensation from other sovereign nations for rain forest preservation in perpetuity, rain forest–rich developing countries will likely continue to burn them down for more short-term, productive use. After all, demand for soya beans and palm oil creates meaningful, immediate income opportunities, and freshly cleared rain forests are perfect for planting fields. Deforestation is, of course, a double whammy. Dense tropical flora are a "natural sink," something we need and want more of. Tropical forests are one of nature's most effective ways of transforming carbon dioxide into oxygen in the air and carbon in the ground. When we burn rain forests down, all their stored carbon gets released. Rather than being a sink, they become a faucet. Double whammy.

Bill Gates shares Project Drawdown's optimism for "new tech." We all should. Gates has pledged to fund any promising new technologies in carbon capture, steel and cement manufacturing, modern agronomy, advanced biofuels, nuclear fusion, and grid efficiencies. These technologies all have the possibility of redressing at least 1 percent of those fifty-one billion tons of carbon when fully brought to scale. One thing everyone should want their retirement funds to be invested in is any scalable technology that meaningfully lowers what Gates refers to as the "green premium." A green premium is the incremental cost between doing business the same old way versus some newly imagined way that promotes net-zero goals. For example, it's been recently estimated that changing America's entire electrical grid to known zero-carbon sources would raise monthly electrical bills by about 15 percent, or $18 a month for an average home. Of course, $18 a month seems manageable, but most of that increase would be borne by lower-income households and those who can least afford it. For some, higher electric bills mean less food. A lower incremental cost would be even better. Reducing the green premiums on steel and cement production are much heavier lifts (no pun intended). "Short of shutting down these parts of the manufacturing

Global greenhouse gas emissions by sector

This is shown for the year 2016 – global greenhouse gas emissions were 49.4 billion tonnes CO_2eq.

Our World in Data

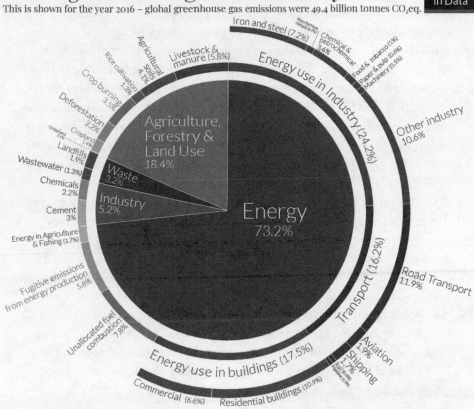

FIGURE 5.3. Charting the (impossible?) path to net zero by industry.
Climate Watch, the World Resources Institute (2020). Licensed under CC-BY by the author Hannah Ritchie (2020).

sector," Gates says, "we can do nothing today to avoid these emissions." As figure 5.3 shows, manufacturing accounts for sixteen billion tons of carbon emissions per year, or about twice the amount generated by transport and home heating and cooling. There is no way to achieve the goals of Paris without fundamentally transforming our approach to heavy industries. Do you have any innovative ideas about how to make steel or cement without the temperature and energy intensity these processes now entail? If so, there's almost certainly a Nobel Prize in your future.

As challenging as all this sounds, there is some reason for hope. For one, dating back to Thomas Malthus in 1798, new technologies have always put the lie to every informed pundit's catastrophic prediction. It turns out human beings are remarkably adept at solving recognized problems far faster and more comprehensively than doomsayers ever expect. Malthus predicted disease, famine, war, and calamity were inevitable because humanity could never create enough food to feed its growing numbers. Innovation and ingenuity proved him wrong. Modern civilization was also supposed to end when Y2K rolled around since no computers were programmed properly for the millennium's end—yet here we are. Every doomsday pessimist to date has been proven wrong. This is not to say climate is not a concern; it clearly is. But claiming humanity cannot survive if temperatures rise another two to four degrees Celsius is simply untrue. The hysteria this claim generates is also unhelpful for developing a range of energy solutions that are sufficiently clean, affordable, safe, and reliable for economic growth to continue.

In fact, it very much appears humanity can continue to make considerable socioeconomic progress whether the goals of Paris are achieved or not. According to almost all estimates for global growth—from the UN and the Organisation for Economic Co-operation and Development to multiple private sector forecasts—GDP will at least double between now and 2050 *even if global policy-makers take no precautionary climate actions*. This data point probably needs to be repeated for emphasis: *even if individuals and governments take no active steps to remediate carbon emissions and temperatures rise significantly, the global economy will still likely double over the next thirty years*. Growing populations generate growth since more human souls require more homes, meals, travel, and health care. In short, even if we don't aggressively shoot for the goals of Paris, it very much seems that good economic times can roll on for a while yet (figure 5.4).

Great, you say. So why spend trillions we don't have to mitigate an increase in global average temperature that doesn't impact our growth? Well, for two reasons. First, growth trajectories are projected to fall dramatically after thirty years in the absence of any action. No action means we could end up living in a "hothouse" world, with no reasonable means for turning the clock or temperatures back.[1] Today, 1 percent of the globe is uninhabitable owing to heat; in a hothouse world, as much as 19 percent could be. Second, if individuals, corporations, and policy-makers actively support a successful green transition, those economic forecasts result in much more cumulative growth. How

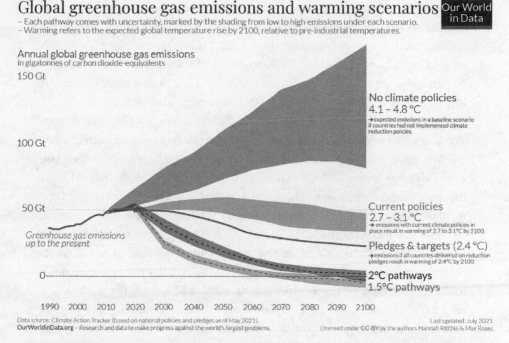

Global greenhouse gas emissions and warming scenarios

- Each pathway comes with uncertainty, marked by the shading from low to high emissions under each scenario.
- Warming refers to the expected global temperature rise by 2100, relative to pre-industrial temperatures.

Our World in Data

Annual global greenhouse gas emissions
in gigatonnes of carbon dioxide-equivalents

150 Gt

100 Gt

50 Gt

Greenhouse gas emissions
up to the present

0

No climate policies
4.1 – 4.8 °C
→ expected emissions in a baseline scenario
if countries had not implemented climate
reduction policies.

Current policies
2.7 – 3.1 °C
→ emissions with current climate policies in
place result in warming of 2.7 to 3.1°C by 2100.

Pledges & targets (2.4 °C)
→ emissions if all countries delivered on reduction
pledges result in warming of 2.4°C by 2100.

2°C pathways
1.5°C pathways

1990 2000 2010 2020 2030 2040 2050 2060 2070 2080 2090 2100

Data source: Climate Action Tracker (based on national policies and pledges as of May 2021).
OurWorldinData.org – Research and data to make progress against the world's largest problems.

Last updated: July 2021.
Licensed under CC-BY by the authors Hannah Ritchie & Max Roser.

FIGURE 5.4. Scenario analysis: plausible carbon and temperature pathways through 2100.
Data source: Climate Action Tracker. OurWorldinData.org. Licensed under CC-BY by the authors Hannah Ritchie & Max Roser.

much more? As much as $25 trillion more through 2040, according to one authoritative report.[2] This incremental growth would come in two primary dimensions. A significant component would come from the extra spending on needed infrastructure and climate preparedness efforts. After all, successful public works translate directly into higher GDP. The rest would come from successfully mitigating damage from more extreme weather. Of course, none of these benefits would be linear. Depending on the steps taken, the composition of future economic growth could and almost certainly would be very different by country, industry, and company within industries. Navigating the pending energy transition will create many national, industrial sector, and individual corporate winners and losers. In extreme scenarios, some countries and many companies would cease to exist.

Take the case of China, for example. Analysts broadly agree China has the most to gain or lose from taking significant action or no action to slow climate change. Its cumulative savings from investing what is needed to successfully transition—a princely sum, estimated to be about 18 percent of the country's total GDP over the next twenty years—would be more than double what it would otherwise lose from taking no action. How much might China lose from taking no environmental action? Perhaps as much as 45 percent of its GDP.[3] China would not remain much of a global force if it lost half of its economy and wealth. No country could or would.

Like other countries, the vast bulk of China's savings would come from averting climate-related damage to existing infrastructure and human capital. You have probably already noticed Beijing is getting a nasty foretaste of the damage that may lie ahead if they do not take climate-related risks more seriously. Crippling sandstorms from the Gobi Desert sparked by China's hotter, dryer temperatures have already started to make China's capital city intermittently unlivable. In March 2021, Beijing experienced its worst sandstorm in more than a decade, with air pollution rising to *160 times* its recommended limit. Flights in and out of the capital were suspended. "It looks like the end of the world," Beijing resident Flora Zou reported. On its present course, Beijing's air quality will almost certainly get worse. In fact, desertification for Beijing now seems a matter of when, not if. Try to estimate the cost of abandoning Beijing's infrastructure and relocating forty million residents to Nanking, China's capital during the Ming dynasty. It would not be cheap.

Beijing is not alone in facing expensive weather-related costs. In 2020, natural disasters globally caused an estimated $210 billion in property damages, the highest amount ever recorded and an increase of 15 percent from an inflation-adjusted average of the last ten years. Between Hurricanes Henri and Ida, 2021 was also pricey. Estimated costs from the 2021 deep freeze in Texas and its impact on the state's grid initially topped $80 billion alone. Many U.S. states and local debt issuers, electric utility companies, and commercial real estate trusts have all been threatened with insolvency since 2010. One—Pacific Gas and Electric, California's largest utility—ultimately entered bankruptcy owing to hundreds of lawsuits from deadly wildfires caused by their downed utility lines. By one estimate, nearly $9 trillion in outstanding debt—more than 10 percent of all bonds outstanding—are now at risk of climate-related downgrades. These problems are not going away anytime soon, either. Climate change could ultimately hike debt servicing costs on corporate and sovereign debt by as much as $300 billion *per year* by the end of this century.

Climate change is and will continue to be incredibly expensive. You now see how an increase in temperature of one degree Celsius could easily equate to a loss of $100 trillion over time. This is why many regulators, central bankers, scientists, and politicians are so alarmed. Humanity is in an unprecedented pickle. The way we have lived and continue to live appears unsustainable. Left unabated, billions of people will be displaced. Weather-related deaths globally have averaged around sixty thousand per year. Spikes to six hundred thousand or one million weather-related deaths per year are possible in the decades to come.

One way to assess the corporate winners and losers from a nonexistent, partial, or complete transition to net zero is to look at the expected costs and savings borne by each industry. This is especially important for investors. It is also the primary analysis driving much innovation behind indexed ESG investment strategies. For example, the cumulative impacts of regulatory and behavioral change, innovation, and supply chain reconfiguration will impact corporations like Microsoft and Edison Electric very differently. Similarly, United Healthcare will be affected more by variable effects from evolving fundamentals and investor preferences driven by climate change more than materials or energy companies like BHP or TotalEnergies of France will. If indexed ESG strategies can successfully navigate these diverging trends, they would be extremely valuable. In effect, this is exactly what most indexed ESG strategies are trying to do.

The aforementioned BlackRock study tries to put some metrics around these differentiating drivers of performance.[4] It estimates the annualized financial returns across information technology and energy companies and finds they could be as much as eight hundred basis points per annum over the next half-decade. If true, choosing the correct climate winners and losers in your retirement fund over just the next five years might impact your investment results by a factor of two. In other words, $10,000 of your invested capital could fall to $7,000 or rise to $14,000. Big difference!

Various ESG investment strategies attempt to navigate these perils and opportunities by intentionally assuming industry tilts toward less threatened sectors and companies in their active and indexed positioning. Adroit investors will experience success, as will their asset managers. The data cited above reflects BlackRock's estimate of how much of the relative outperformance expected is based upon a "repricing channel." It's effectively an increase or decrease in their earnings multiple—reflecting expectations for future growth—versus how much results from a "fundamentals channel," meaning actual earnings gained or lost. Total under- or outperformance is the net of these two channels added together. Information technology and health care companies are much better positioned than utility and energy companies because they will have more predictable *and* higher earnings. More predictable and higher earnings are invariably rewarded over time with higher investment multiples. Indexed equity products that screen out riskier industries and overweight those less exposed will ride these repricings axiomatically, should they occur. If these earnings assumptions and repricing forecasts are correct, many smart indexed ESG strategies would work handsomely.

Just as there is measurable climate risk between industry sectors, there will also be outperformers and underperformers within industry sectors and within companies in the same industry. Consider the oil majors. The global economy will undoubtedly need some traditional oil products and services from the likes of Chevron, Exxon Mobil, Saudi Aramco, Shell, TotalEnergies, and other energy conglomerates for generations to come, if not in perpetuity. There is no imaginable future without some carbon-based energy consumption. But even among these names, one or two will surely meet the challenge of powering our future more sustainably and strategically than others. Which ones? I don't know—but I sure hope my retirement assets will be managed by those who do.

The Schumpeterian forces of creative destruction will prove no less potent in the energy sector than they have been in the automobile sector. Indeed, creative destruction will prove even more critical in the energy space, given the massive transition that is now taking place and is set to continue. A great deal of that under- and outperformance is apparent already. For example, the index giant Morgan Stanley Capital International (MSCI) estimates that the cumulative difference between the most and least prepared energy companies from 2013 to 2020 in earnings per share was more than 75 percent. Their share price performance differential over that same period was about 25 percent. These are huge differentials. Back to our $10,000 retirement savings example: successful positioning means your account could be worth $12,500 or $7,500. Yikes! Many screened indexed products fail to capture differentials like these because they absent themselves from entire sectors. When you ban *all* energy sector investments, you effectively absent yourself from the opportunity to pick more innovative companies in a vital industrial sector that will determine whether and how net zero is achieved. While indexed strategies have the advantage of being cheaper and approximately correct, they also have more limited chances of capturing idiosyncratic alpha opportunities in an era in which such opportunities will almost certainly prove rife. To put it more succinctly: many indexed ESG strategies will either be approximately correct or precisely wrong.

Probabilities are an inherent component of every investment strategy. For example, a decision to favor equities over bonds at various points in the economic cycle is based upon the assumption that corporate earnings are likely to expand owing to accelerating economic growth and that interest rates are likely to rise as they do. As bond prices go in the opposite direction of bond yields, you are almost always better off owning more stocks and fewer bonds while economic expansions are beginning to take hold, and more bonds and fewer stocks right before recessions set in.

When it comes to pricing in probabilities of climate policies being enacted and factoring in how those probabilities might impact specific asset classes like equities or bonds, however, a wide array of assumptions ultimately drives an even more comprehensive array of potential outcomes. For example, the report from BlackRock referred to earlier projects that European equities could return close to 9 percent per year with good climate policies and 7 percent per year without—but they acknowledge a band of uncertainty around these estimates of nearly 20 percent. Interestingly, in the case of China, annual expected equity returns with or without climate action are effectively projected to be a wash, or about 5 percent per annum. That said, BlackRock acknowledges that interquartile performance could be as much as −8 percent to +17 percent—a whopping swing of one-quarter of the total capital at risk! So even after putting politics aside—which you shouldn't do—investing in China over the decades to come is not for the faint of heart. The traditional investment axiom holds: for outsized reward, one must be willing to assume some serious outsized risks.

Similarly, employing options strategies against one's stock positions—puts or calls, which allow you to buy or sell individual stocks at set prices over set periods—are all based upon probabilistic pricing models and the time value of money. If you want to have the privilege of buying or selling a stock in the future, you must pay an "option premium" that reflects the market value of that specific event coming to pass. Many investors like to sell the right to buy stocks they own to someone else to generate some extra income, a strategy known as "covered call writing." If the stock goes down, they merely keep the premium income they received. If the stock goes up, they keep the premium income but are also obligated to sell the stock at the strike price they agreed to.

From a purely financial insurance perspective, more than anything, the world today needs a huge put option on itself. Ideally, pensioners and investors would be able to buy insurance to protect their portfolios against the worst-case scenarios of physical damage from climate change. But who could possibly provide such insurance, and at what price? While insurance companies still underwrite certain types of risk relating to climate damage, comprehensive catastrophic coverage has become too expensive in many locations. There is a good reason for this: the growing prospect of total devastation involves costs that no private insurer could sensibly bear and that no public insurer could sensibly provide.

So far in this chapter, we have focused our discussions of climate risk on one primary dimension: physical risk. However, when assessing the financial or investment costs of climate risk, at least three other dimensions come into play. The first of these is regulatory risk.

Today, 130 governments around the globe have either implemented or are strongly considering implementing rules and regulations to accelerate their transition to a lower-carbon future. The United States rejoined the Paris Agreement shortly after President Joe Biden moved into the White House, reuniting with the European Union, China, Japan, South Korea, and 190 other nations. Collectively, these nations account for more than 90 percent of global emissions. More than 2,100 laws have already been proposed by these nations, including penalties and incentives like carbon taxes and sustainable tax credits. COP 26 in Glasgow in 2021 resulted in additional commitments to lower methane emissions and "phase down" other more carbon-intensive electricity sources, like coal. The cumulative effects of these policy actions could have meaningful impacts on equity and asset valuations in every country and industry. Energy companies now report more than $1 trillion of their enterprise value comes from proven oil reserves; they think of their proven oil in the ground like money in a bank. The problem is that these listed assets could quickly become less valuable or even "stranded" owing to the policy measures mentioned earlier. As more and more companies and communities develop plans to stop using fossil fuels and aim for net-zero carbon emissions in the years ahead, lots of grounded oil may remain there. Writing down the values of some of those energy assets could prove very material. Some energy assets could even switch sides on a corporate balance sheet; instead of being assets, they could become liabilities.

A second, more positive risk extends from all the innovation that emerging technologies engender. While fossil fuels currently provide more than

80 percent of the globe's power needs, renewables are projected to provide as much as half over the next fifty years. Clean energy patents are on the rise, accelerating the probability that some companies will leapfrog existing suppliers, rendering their business models and physical assets obsolete. Photovoltaic cell production costs fell 89 percent between 2009 and 2019, massively reducing the green premium between solar and other electricity options. This is positive news for consumers, greenhouse gas emissions, and the planet. However, it is less good if you are an expensive producer of those cells or a company or coal miner who provided raw materials for energy production for decades. Damn it, there's old Schumpeter again!

Since 2007 the number of companies that have begun to report reliable data on their energy use and carbon footprints has grown sevenfold, to nearly ten thousand. This growth highlights the last potential transmission mechanism of climate risk to investment risk. In addition to physical, regulatory, and innovation risks, there are *reputational* risks. Reputational risks readily translate into consumer preferences, lost client loyalties, and vulnerable revenues.

A logical response by global consumers to historic heat waves, unprecedented floods, and raging wildfires that indelibly impact human lives may well be to single out and penalize proven contributors to climate change. Over time, companies that successfully develop solutions for our mounting climate challenges will likely win consumer favor and market share. Of course, the converse is also true: mindless polluters could find themselves quickly shunned by consumers and financially reprimanded by regulators. For example, Volkswagen's shares dropped 37 percent in the days immediately following revelations that they had fabricated emissions data for their diesel vehicles. As a result, Volkswagen had to pay more than €27 billion in fines and watch their market share drop for two straight years before stabilizing. An era of greater consumer awareness and more corporate disclosures heightens the probability of other episodes like Volkswagen's coming to light.

The International Renewable Energy Agency—an intergovernmental organization based in Abu Dhabi—estimates the world will need to invest $115 trillion through 2050 just for cleaner technologies and more sustainable solutions. They further estimate such a shift could create three times more jobs than it destroys. Why? It turns out renewable energy industries are more labor intensive than those involving fossil fuels. Based in Geneva, Switzerland, the International Labour Organization also estimated similar job gains. In 2018, they claimed a shift to more sustainable solutions would eliminate six million jobs but create twenty-four million. Their analysis suggests the energy transition could resemble previous tectonic shifts in human behavior—like the move to industrialization in the seventeenth and eighteenth centuries or the more recent digital age. Of course, all of this sounds encouraging at a macro level. How it sounds if you are a coal miner in West Virginia or a wildcatter in the Permian Basin is undoubtedly different.

So, let's quickly recap. Climate risk is investment risk. Climate risk is also country risk, physical plant and equipment risk, enterprise risk, and reputational risk. Oh, yes, and climate risk is here to stay.

With trillions of dollars in motion, however, climate risk is also an investment *opportunity*—in other words, a chance to take advantage of favorable new trends and emerging solutions, separating winners from losers. This means getting the "E" within ESG investing right will be essential for successful risk mitigation and alpha generation. Big money will be made and lost. In fact, getting the "E" in ESG investing right is so important that, in many cases, it will overwhelm other "S" and "G" considerations. Given that optimizing risk-adjusted returns is so crucial, it's evident that thoughtful investors and financial advisers must segment "E," "S," and "G" risks and opportunities very deliberately.

Think of it this way. Many people like sweet, sour, acidic, and pungent spices—but they would never put them all into the same dish. One flavor or fragrance would overwhelm all the others. Putting everything together means none would be appreciated for what they are.

Owning companies with excellent "S" and "G" but problematic "E" scores, all mixed together, may result in the same problem: the final dish may be unpalatable. Averaging these three variables often masks how potentially devastating specific "E" exposures may be, moreover. Similarly, debilitating "G" concerns could overwhelm positive "E" and "S" attributes. If a corporation uses fake accounting or engages in chronic insider dealing—two classic "G" failures—stellar "E" and "S" credentials won't matter one whit. Regrettably, this is precisely what some asset managers and investment advisers are doing when they say their funds are "ESG integrated." They base their investment decisions on aggregated ESG scores that may well fail to account for the disqualifying nature of specific "E," "S," or "G" considerations.

While it is very early going and we have much more ground to cover on ESG, one important revelation already seems to be clear. If investors, regulators, activists, and academics want ESG investing to promote transparency and the interests of all stakeholders, it's increasingly clear that marital counseling may not be enough. In the interests of all sides, it may be best for "E," "S," and "G" to divorce. Their relationship appears irreconcilable. If mutual understanding and clarity are important, it's probably in the best interests of all concerned that they go their separate ways.

But rather than gradually correcting course and sorting themselves out, flows into investment products based upon tilting and timing composite ESG scores are accelerating. They are growing in no small part because people believe implicitly that they are useful vehicles for saving the planet and remediating social and environmental ills. ESG flows are also growing because many regulators and influencers say they should be.

And leading the official bandwagon for more widespread ESG adoption is the largest international organization in history: the United Nations, backed by its 193 member states.

Chapter Six

WHAT'S THE UNITED NATIONS GOT TO DO WITH IT?

We the peoples of the United Nations vow to employ
international machinery for the promotion of the economic
and social advancement of all.

FROM THE PREAMBLE OF THE CHARTER OF
THE UNITED NATIONS

We are getting closer to the technical deep dive on ESG I promised in the introduction, but it turns out we can't talk about ESG investing—how it came into being, what it's trying to do, how it might be strengthened and optimized, and the caveats it presents—without first talking about the United Nations. And it turns out we can't talk about the United Nations without first examining other historic efforts for global collaboration and cooperation. In short, we need to look back before we look forward. Way back.

Global cooperation is incredibly hard. It's also remarkably uncommon. Nations almost always look at others as competitors or enemies, not collaborators. They do so even when they have shared goals. "The nation will continue to be a central pole of identification, even if more and more nations come to share common economic and political forms of organization," the leading political scientist Francis Fukuyama claims.[1] If Fukuyama is right, nations seem destined to bicker, or worse as we now see in Eastern Europe.

That said, every so often—no more than once or twice a century, tops—humanity's collective conscience appears to coalesce around a few core principles and bold ideas to promote peace, security, and human progress. On even fewer of those occasions, a group of aligned and enlightened nations formally bind together, promising to promulgate and defend those principles and goals. The Peace of Westphalia, which ended the Thirty Years' War in 1648, is one example. The *Pax Mongolica*, which brought relative stability to

Eurasia in the thirteenth and fourteenth centuries, after the death of Genghis Khan, may be another.

The United Nations Universal Declaration of Human Rights (UDHR) is almost certainly the most monumental and consequential example of this kind of collective moral commitment in modern times.[2] Drafted in a unique moment of moral rectitude forged by the most lethal and destructive war humankind had ever inflicted upon itself, victors and vanquished vowed not only "never again" but also to achieve far more inclusive economic and social flourishing than humanity had ever known. On December 10, 1948, only three years after the United Nations itself was created, the UDHR was formally adopted by the still-in-formation fifty-eight-member UN General Assembly.[3] The UDHR defined an agenda for human flourishing that was audacious in scope. It has effectively operated as the UN's social mission statement ever since.

The declaration's drafting committee was chaired by the formidable Eleanor Roosevelt (figure 6.1). I'm not sure how much you know about Eleanor

FIGURE 6.1. Eleanor Roosevelt and the Universal Declaration of Human Rights.

Roosevelt, but I'm willing to wager it's not enough. Ms. Roosevelt ranks among history's more underappreciated heroines. As the first presidential spouse to hold regular press conferences, host weekly radio shows, and write daily newspaper columns, Ms. Roosevelt set multiple precedents and standards. In so doing, she taught every aspiring professional—male and female— how to maximize one's personal impact through immutable principles and tireless pursuit. To these traits she added grace and elegance. As is too often the case with famous men, much of FDR's considerable historic legacy actually belongs to his wife, Eleanor.

The declaration begins with a cornerstone tenet: the universality of human dignity. It is a powerful first assertion, one shared with other long-standing traditions, including Catholic Social Teaching. The implications of universal human dignity are profound. The UDHR claims that every human being has the inherent right to be valued and respected for their own sake, a right that has been entrusted to them by divine inspiration owing to their birth. This dignity is inherent. It is not given by man. Birth alone begets human dignity. Any attempt to negate it or take it away is a gross violation of nature's order. The declaration goes on to enumerate a seemingly uncontentious litany of additional inalienable rights, among them life, liberty, security, protection of law, freedom of expression, and property. It then turns to a number of loftier socioeconomic goals. Every human being, the declaration reads, is entitled to employment, social security, equal pay for equal work, the right to rest and leisure, a living wage, free education, and full artistic expression. Those living wages must be enough for housing, food, clothing, and medical care, as well as a comfortable retirement.

Every human being—not just a favored few.

Given Ms. Roosevelt's moral presence and powers of persuasion, there were almost no disagreements about the declaration's fundamental principles. Remarkably, there were also relatively few debates about economic end goals. In fact, no country or delegation vigorously contested the universality of human dignity or how a life well lived might look. It's hard to imagine why one would. The vision for how all people should strive to live in harmony and shared abundance as envisioned by the UDHR was intentionally utopic. You'll recall that Utopia is the perfect, though imaginary, world Saint Thomas More first wrote about in 1516. It comes from the Greek *outopía*, which literally translates to "no place." The world the UDHR envisions and calls for has never existed. It did not exist when the UDHR was written. It exists nowhere on Earth today.

Unsurprisingly, the debate became more heated among delegates once the discussion shifted to how signatories were to translate the declaration's utopic goals into practice—moving from "ends" to "means," in other words. Poorer countries, led by Egypt and India, rightly bemoaned their evident shortage of material resources and inability to enjoy the kinds of widespread prosperity that the declaration called for. This was not for lack of desire, of course. Poorer nations wanted their peoples to live just like those in wealthier countries in all the manners called for in the declaration. Without the means, however, the declaration's objectives seemed impractical, a collection of unattainable aspirations rather than specific, practical policy objectives. In 1948, no country—wealthy or poor—had a viable *pathway* to enact all the declaration's ideals, let alone proximate attainment. The goals of the UDHR were not possible even with more equal distribution of existing resources. The types of social infrastructure, ubiquitous leisure, fulfilling work, and retirement comforts the declaration dreamed of were not possible in the United States in 1948, even though America was then the richest and strongest nation on Earth. These universal gaps and explicit ideals ultimately meant the declaration was intended to be aspirational in nature, an idyllic vision of what was hoped for rather than what might at any specific point in time practicably be.

Beyond adequate financial means, however, there were even more vociferous disagreements about the philosophies and methods needed to achieve greater wealth and well-being, particularly with reference to political structures. The Soviet Union objected to nearly every tenet on civil rights, repeatedly vetoing any endorsement of democracy or democratic principles. Remarkably, however, these differences did not doom the declaration. In the end, forty-eight of the fifty-eight UN General Assembly members voted in favor of the resolution. Nine members abstained or declined to vote. Abstentions included five Soviet-aligned nations plus Honduras, Saudi Arabia, South Africa, Yemen, and, of course, the Soviet Union.[4] Then as now, it was unclear whether the leaders of the Soviet Union (or Russia) ever cared about any nation other than their own.

The global economy experienced regional pockets of growth during the 1960s and 1970s. Regional wars precluded progress across parts of Asia, South America, and Africa, and aggressive actions by the nations of the Organization of the Petroleum Exporting Countries (OPEC) painfully penalized commodity importers in 1973 and 1974. Another oil shock took place in 1979, in the wake of the Iranian Revolution. Global growth attained a more consistent and inclusive pattern of expansion during the 1980s and 1990s. Economic liberalization and the broadening of cross-border trade—a process now referred to as "globalization"—gathered pace after the election of President Ronald Reagan. It accelerated up to and through the presidency of Bill Clinton, illustrating that neither the Republican nor the Democratic Party has an exclusive grip on salient, market-based policies. Market liberalization and globalization meant that more countries experienced improved growth outcomes simultaneously. Asia—more specifically, China—was a central part of this expanding growth dynamic. After composing only 2.8 percent of global GDP in 1980, China's global share more than doubled by the year 2000 and quintupled by 2013 to more than 15 percent. As global living standards broadly improved, more of the aspirational economic and social goals enumerated in the UDHR came within reach of more nations. Rising wealth brought social advancements, improved health care, and better education. Rapidly declining poverty rates across much of the globe increased global consumption, spawning a virtuous circle of rising incomes. At the same time, it was becoming increasingly apparent that unprecedented global economic growth was degrading the environment and further exacerbating income gaps *within* developed and developing nations.[5] The same fissures that preoccupy many activists today were already apparent decades ago.

It is against this broadly positive economic backdrop and the momentous occasion of a new millennium that the United Nations assembled the largest gathering of national leaders in history. In early September 2000, one hundred heads of state, forty-seven heads of government, and three monarchs joined eight thousand other ministers and diplomats at the UN's global headquarters in New York City to discuss and ultimately approve an ambitious collection of Millennium Development Goals. More than 150 countries participated, a stunning feat. In the end, the United Nations Millennium Declaration reaffirmed many of the values and principles of the UDHR. To that list they added respect for nature and disarmament, as well as a chapter devoted to the unique needs of Africa. Signatories also committed themselves to eight specific targets to be met by 2015: (1) the eradication of poverty and hunger; (2) universal primary education; (3) greater gender equality; (4) reduced child mortality; (5) improved maternal health; (6) the end of HIV and AIDS, malaria, and other diseases; (7) environmental sustainability; and (8) a global partnership for development. It is important to note that none of these goals had previously been within reach; only the economic expansion that two decades of globalization and free markets had enabled made them possible. Unlike the UDHR, moreover, specific metrics and timetables were laid out. Each goal was assigned measurable targets and dates by which quantifiable milestones were to be achieved. Fifty-two years after its adoption, the UDHR was moving from theory to practice, from aspiration to implementation. Unprecedented global growth meant many of the economic dreams in the utopic UDHR could finally become uncontested realities. Economic growth underwrote social progress.

From 2000 to 2010, significant progress continued to be made on each of these goals. So much so, in fact, that some goals were achieved in several categories well ahead of schedule. These included much lower global poverty rates, reduced child mortality, and higher literacy rates. At a UN conference on sustainable development in Rio de Janeiro in 2012, some elements of the Millennium Declaration were even deemed insufficiently ambitious. Unpredicted and unprecedented progress led many UN members to believe the time had come to set even greater human development goals. Social and economic progress was accelerating like never before! Over the next three years, the UN debated and ultimately adopted a total of seventeen Sustainable Development Goals (SDGs), all of which were designed to be achieved by 2030 (figure 6.2). Many—like eliminating poverty and hunger, promoting good health and well-being, and providing better education—harkened back to the UDHR.

FIGURE 6.2. The UN's seventeen Sustainable Development Goals.
https://www.un.org/sustainabledevelopment/. The content of this publication has not been approved by the United Nations and does not reflect the views of the United Nations or its officials or Member States.

To these, six new objectives were added: (1) clean water; (2) affordable and clean energy; (3) responsible consumption and production; (4) reduced inequality; (5) sustainable cities and communities; and (6) climate action. None of these had been envisioned in 1948 because none of their underlying causes and concerns were then apparent. Environmental priorities came last to the SDGs because they last manifested themselves in the public mind.

Incidentally, it is relatively easy to map all seventeen SDGs back to our North Star, namely, more inclusive and more sustainable economic growth. SDGs 1, 2, 10, 12, and 16 all relate to inclusivity; SDGs 3, 4, 5, 6, 11, 13, 14, and 15 all involve sustainability; and SDGs 7, 8, 9, and 17 concern growth and prosperity.

So, you must be asking by now, What does *any* of this have to do with ESG investing? Let me tell you.

In 2005, as an integrated part of all these development initiatives, then UN Secretary-General Kofi Annan invited a couple dozen international investment firms and financial organizations to consider forming a series of principles for investment that could help accelerate SDG attainment. Annan's investor group was drawn from institutions across twelve countries and was further supported by a seventy-person group of experts from the investment industry, intergovernmental organizations, and civil society. Among this founding group were representatives of several of the world's largest pension plans from the United States, Europe, Africa, and Asia; four religious superannuation funds; and a handful of banks from Europe and Japan. This advisory group took it upon themselves to define what they considered to be finance's direct responsibility to promote the economic and social ends of the UDHR and the SDGs that ultimately grew out of them. They launched their Principles for Responsible Investment (PRI) in April 2006 at the New York Stock Exchange. Secretary-General Annan,[6] himself, rang the bell.

Today, the UN-supported PRI principles form the theoretical basis for virtually all global initiatives relating to ESG investing. The PRI assists its members in planning how to incorporate ESG factors into their decision-making. Just like the UDHR and the SDGs, the PRI principles are aspirational and voluntary. Signatories pledge to follow the principles but are not bound to any specific action or auditable process.

THE PRI's SIX PRINCIPLES FOR RESPONSIBLE INVESTMENT

- **Principle 1:** We will incorporate ESG issues into investment analysis and decision-making processes.

- **Principle 2:** We will be active owners and incorporate ESG issues into our owner-ship policies and practices.
- **Principle 3:** We will seek appropriate disclosure on ESG issues by the entities in which we invest.
- **Principle 4:** We will promote acceptance and implementation of the Principles within the investment industry.
- **Principle 5:** We will work together to enhance our effectiveness in implementing the Principles.
- **Principle 6:** We will each report on our activities and progress towards imple-menting the Principles.

As of December 2021, more than 3,700 investment managers and 650 asset owners from eighty-six countries were signatories to the PRI. The total rises to more than seven thousand institutions when you add in all the other corporate commitments. In aggregate, these institutions oversee more than $120 trillion in assets. This makes the PRI the largest agency in the world dedicated to responsible investment and the development of ESG invest-ment techniques and principles. It is supranational, by which I mean it is not beholden to any government or agency. In fact, it is not even a formal part of the UN, even though it is nominally supported by it. The PRI is run by a separate board elected by its member signatories. The PRI's board pledges to work in the long-term interests of its signatories, the financial markets and global economy more broadly, and ultimately of the environment and society as a whole.[7]

"Investors have a critical role to play in addressing all the issues the world faces," the PRI website proclaims in a video introducing its master Blueprint for Responsible Investment. The $120 trillion in assets signatories oversee "can be used to support sustainable projects now, while also providing a dig-nified retirement for all in the future. Our aim is to create responsible inves-tors, sustainable markets and a prosperous world for all." In other words, more inclusive and sustainable growth.

But the PRI simultaneously admits it cannot achieve any of its objectives without a lot of help from many others.

"We can't do this alone," their video goes on to assert. "It will only be when investing responsibly becomes a way of life for everyone that we can produce a prosperous world for all." As the PRI's pedagogical tools delve deeper into how each principle can be applied, they also leave broad latitude for interpreting exactly how to put each into practice. Detailed descriptions of possible actions are given for each principle. For example, under Principle 1, the PRI website lists seven possible actions signatories may adopt.

PRINCIPLE 1: WE WILL INCORPORATE ESG ISSUES
INTO INVESTMENT ANALYSIS AND DECISION
-MAKING PROCESSES.

POSSIBLE ACTIONS:

- Address ESG issues in investment policy statements.
- Support development of ESG-related tools, metrics, and analyses.
- Assess the capabilities of internal investment managers to incorporate ESG issues.
- Assess the capabilities of external investment managers to incorporate ESG issues.
- Ask investment service providers (such as financial analysts, consultants, brokers, research firms, or rating companies) to integrate ESG factors into evolving research and analysis.
- Encourage academic and other research on this theme.
- Advocate ESG training for investment professionals.[8]

In other words, like many self-regulatory bodies, the PRI exists to help its members promote best practices as they themselves come to define those practices over time. Its members recognize that these practices will evolve and that they must be open to learning from one another. They make no effort to codify principles into binding regulatory rules. The PRI is not a policing body. Rather, it exists for the benefit of its members, all of whom have vowed to use financial tools and investment practices to help attain the SDGs. In this respect, they are rather like the World Association of Chefs' Societies, though focused on creating optimal financial and investment outcomes rather than global culinary standards.

The PRI defines responsible investment as a strategy or practice that incorporates environmental, social, and governance factors into investment decisions and active ownership. As shown in figure 6.3, the PRI has enumerated fifteen examples of material ESG issues. These are not intended to be exhaustive, nor are they ordinally ranked. They are illustrative and open to further improvement and more sophisticated iteration.

In principles 1 and 2, and repeatedly in documents and briefings supplied by the PRI, you'll find a curious technical term used over and over: *material*. Signatories have pledged to incorporate "material ESG issues" into all of their

ENVIRONMENTAL

- climate change
- resource depletion
- waste
- pollution
- deforestation

SOCIAL

- human rights
- modern slavery
- child labour
- working conditions
- employee relations

GOVERNANCE

- bribery and corruption
- executive pay
- board diversity and structure
- political lobbying and donations
- tax strategy

FIGURE 6.3. ESG concerns enumerated by the PRI.
Reprinted with permission from UN-Supported Principles for Responsible Investment (PRI).
https://www.unpri.org/an-introduction-to-responsible-investment/what-is-responsible-investment/4780.
article; https://www.unpri.org/pri/what-are-the-principles-for-responsible-investment.

investment analysis, decision-making, ownership policies, and practices. In figure 6.3, the PRI enumerates fifteen material concerns that bear watching.

It's quite clear that material ESG issues are important. In fact, they are the primary variable behind all ESG risk mitigation and asset allocation. That's why we are about to spend an entire chapter discussing them.

Chapter Seven

MATERIALITY

Not everything that counts can be counted, and not
everything that can be counted counts.
ALBERT EINSTEIN

If you torture the data long enough, it will confess to anything.
RONALD COASE

On July 24, 2019, the Federal Trade Commission imposed a record-breaking $5 billion fine on Facebook[1] for doing something few other corporations could even imagine: violating their customers' privacy. This sum was twenty times greater than the most significant privacy or data security penalty ever imposed and one of the most severe penalties ever assessed by the U.S. federal government on any corporation for any violation.[2] What the hell did Mark Zuckerberg and his colleagues do to earn such opprobrium?

Something material.

Facebook's most egregious transgressions took place over several critical months in 2016. That's when British voters narrowly backed Brexit 51.9 percent to 48.1 percent. A few months later, Donald Trump shocked the world by capturing Wisconsin, Michigan, and Pennsylvania by less than 1 percent of the total votes cast, snatching what had seemed a sure victory away from Hillary Clinton. Facebook's scandal came to light only in the spring of 2018, when a group of ex-Cambridge Analytica whistleblowers revealed their firm had systematically harvested personal data from nearly ninety million Facebook users. Cambridge Analytica used this personal data to target highly charged political ads and spread misinformation relating to the 2016 U.S. presidential and UK Brexit campaigns. As outlandish as it sounds, Facebook's transgressions ostensibly put Donald Trump in the White House and precipitated the most destabilizing political event in the history of the European Union. Five billion dollars hardly seems sufficient given the potential significance of these events. In the weeks immediately following the Cambridge Analytica revelations, Facebook shares plunged by more than 43 percent, eviscerating hundreds of billions of dollars of shareholder value. Tens of millions of Facebook clients summarily closed their accounts, including me.

The concept of materiality as it relates to the UN's Principles for Responsible Investment involves financial risk. An ESG risk is not considered material unless it has a tangible financial consequence. Material ESG risks aren't about morals, in other words: they are about money.

Many chest-thumping analysts claim they saw Facebook's issues coming. One early signal flashed red in 2010 when the founder of Facebook, Mark Zuckerberg, gave an off-color speech about how the global age of privacy was coming to an end because people seemed increasingly keen to share more data about themselves publicly. His implicit endorsement of emerging platforms like Twitter conveniently coincided with Facebook's desire to grow revenues from sponsored content and page views. His congressional testimony nine years later confirmed his catastrophic misjudgment: "It's clear now that we didn't do enough to prevent our tools from being used for harm. That goes for fake news, foreign interference in elections, and hate speech, as well as developers and data privacy." Prudence now seems to govern Facebook's privacy and data-sharing pronouncements. But behind their public caution lie material financial concerns. In a 2021 update to their more than three billion users, Facebook stated, "We're not asking for new rights to collect, use or share your data on Facebook. We're also not changing any of the privacy choices you've made in the past." Apparently, a $5 billion fine and a 43 percent share free fall helped clarify their strategic thinking.

The materiality of data privacy matters much more to social media groups and banking firms than it does to other industries, such as beverage companies or mining firms. But mining companies—like Massey Energy—have other, equally material worries. Massey Energy's unique vulnerabilities came into painful view on April 5, 2010, when an explosion occurred at their Upper Big Branch coal mine in West Virginia, killing twenty-nine of their thirty-one on-site miners. This tragic yet entirely avoidable event was the deadliest mining accident in the United States since 1970. Remarkably, the federal regulator Mine Safety and Health Administration had issued more than *three thousand* safety violations against Massey Energy between 1995 and 2010, but no rule required disclosing any of these summations. Massey had been a serial violator of worker safety for years but addressed its failures only after the unspeakable Upper Big Branch tragedy struck. As a result,

Alpha Natural Resources agreed to purchase the company and assume all its liabilities in 2011. Four long years later, Don Blankenship—the CEO of Massey at the time of the explosion—was sentenced to one year in prison for willfully conspiring to violate federally regulated safety standards.

Facebook's and Massey's catastrophic setbacks could hardly be different in cause, but they share essential commonalities. Both corporations had enterprise-threatening risks embedded in their business models—but those risks lurked below the surface. In Massey's case, they were legal but irresponsible, since disclosing their three thousand safety violations had not been required. In Facebook's case, they were both illegal and irresponsible. Because these risks were unknown to the broader public, they could not be appropriately discounted in their enterprise value. The ESG disclosure movement's key objective is to unmask all types of material risks systematically before they manifest themselves. Since investors ultimately bear the cost of wrongdoing, material risk equates to their financial vulnerability. Forcing greater transparency about market-moving environmental, social, and governance concerns would likely incline executive leaders and their boards to mitigate those vulnerabilities wherever and whenever possible. Proper disclosures would also help investors be more discerning, valuing heretofore unseen and emerging risks more appropriately in those firms' share and bond prices. For investors, more transparency is always good. Transparency allows discerning market players to price the probability of a broader range of possible outcomes more accurately.

But it's also easy to see many ways improved corporate disclosure procedures could still break down. What is financially material to one person, place, or industry at any given point in time may not be to another. How one weighs data security versus public safety by company and industry is also less than obvious; indeed, it's difficult to imagine it could ever be entirely objective. As we now know, public opinion about what is material also evolves. For example, a focus on management diversity and employee inclusion practices has risen demonstrably in just the last few years, after years of sullen indifference. As the societal focus on diversity, equity, and inclusion—DEI, in today's parlance—has intensified, material financial concerns relating to that focus have risen concomitantly. DEI issues became materially risky relatively recently, something they hadn't been before. Social concerns evolve. This means "S" risks evolve with them.

Most importantly, actionable material disclosures depend crucially upon what corporations choose to report. To begin with, the integrity of the information provided is vital. Lamentably and remarkably, many firms intentionally lie—as Volkswagen did in their extraordinary, prolonged efforts to misrepresent emissions from their diesel vehicles. Others tell the truth, just not the whole truth—as in the 2018 promotional video of Nikola's electric truck that reportedly generated zero emissions. (Note: In the video, it was coasting down a hill—and yes, gravity is carbon free!) Of equal significance is the consistency of the data between firms and industries. For example, reporting self-assessed employee satisfaction scores is highly subjective, dependent upon how those surveys are conducted and which questions are asked. Unless and until the same methodologies are used by and among peers, comparable metrics relating to employee satisfaction and many other ESG categories may prove more misleading than illuminating. How all reported data is ultimately aggregated and reported is equally consequential. Should one use a numeric scoring system or color-coding to convey relative materiality risk? If colors, what exactly do, say, red, yellow, and green mean? Finally, regulatory bodies and accounting standards in different jurisdictions are at very different stages of their disclosure journeys. When reporting standards are more demanding, and the information revealed more damning, in one country than in another, companies domiciled in those jurisdictions may become disadvantaged. Perversely, honest and well-intended disclosures might make well-intended, transparent companies more vulnerable.

The risks of jurisdictional disparities unfairly penalizing the best actors is not a theoretical concern. As it so happens, Europe excels at detailed regulation, a natural response perhaps to compensate for its relative surfeit of technological and industrial innovation.[3] America is laxer, and Asia badly lags. If the ESG reporting noose continues to tighten more in Europe than in America, corporations may well choose to reincorporate in the United States, taking advantage of a more accommodating environment. This is what happened with corporate taxes, after all, and more recently with Shell, PLC.[4] Corporations naturally gravitate to more welcoming jurisdictions. From a shareholder perspective, it would be irresponsible not to.

All three primary variables—corporate disclosures, related aggregation techniques, and regulatory requirements—are ever-evolving. This means ESG optimization involves dynamic, multivariable calculus. Regulators and corporate executives are solving for multiple ends simultaneously. There will always be trade-offs. Regulatory trends driven by climate and social justice concerns are themselves constantly changing. Corporations must both comply with existing social norms and regulations while anticipating new ones.

The largest provider of equity indices, Morgan Stanley Capital International (MSCI), oversees what many consider the most extensive research operation quantifying material risks relating to environment, social, and governance concerns. MSCI's primary objective is to give investors informed choices that accurately reflect the precise exposures they seek. The idea is not to drive decisions but to enable informed choice. In their unique ESG ranking methodology, environmental and social concerns are segmented into four broad categories. They further break down their governance concerns into two groups. Each category has up to six subcategories, meaning MSCI strives to measure a total of thirty-five specific sources of potential material risk. Figure 7.1 illustrates MSCI's ESG materiality risk spectrum. From this, you can readily deduce that it's a pretty sophisticated and detailed effort.

It is also complicated. This particular figure happens to spotlight the most financially material ESG concerns in one of MSCI's industry categories: soft drinks. When you look at the gray boxes with dashed outlines, therefore, try

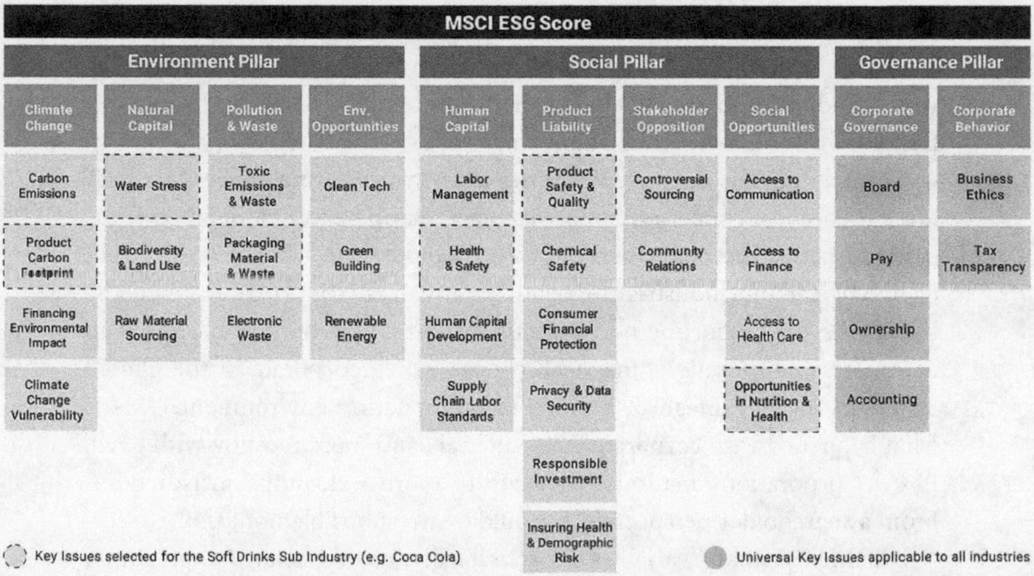

FIGURE 7.1. Dozens of distinct material considerations lie behind MSCI's ESG scoring methodology.

Reproduced by permission of MSCI ESG Research LLC ©2022; all rights reserved. Although Terrence Keeley's and Columbia University Press's information providers (including MSCI ESG Research LLC and its affiliates, the "ESG Parties") obtain information from sources they consider reliable, none of the ESG Parties warrants or guarantees the originality, accuracy, and/or completeness of any data herein and expressly disclaim all express or implied warranties, nor shall any of the ESG Parties have liability for any errors or omissions in connection with any data herein, or liability for any damages.

to think Coca-Cola or Pepsi. According to MSCI, less than one-third of their identified potential sources of material environmental and social risks pertain to corporations in the soft drinks industry. These include the product's carbon footprint, water stress, packing material, and waste; worker health and safety; product safety and quality; and nutrition and health. In addition to these, MSCI regards all six governance considerations to be financially material to soft drink firms like Coca-Cola or Pepsi, including business ethics, tax transparency, accounting, board composition, pay, and ownership. In MSCI's unique calculus, each of these factors poses material financial risks to every soft drink firm's operations and enterprise value. None of the other risk categories is considered material by MSCI. One can see why this would be so, moreover. Data security and climate change vulnerabilities do not matter materially to Pepsi and Coke, at least not as they do to Facebook or Massey Energy. For every industry and company MSCI rates, they focus only on those material risks that genuinely matter, no others.

MSCI takes all the data companies report as a primary input to their ESG rating model, giving heavier weight to the most standardized data sources across the industry. To this, they add other data sources as well as their qualitative judgments. According to their website, these unreported sources and qualitative considerations are crucial, often amounting to half of their final assessment:

> For the average company, self-reported ESG information comprises approximately 50 percent of the full suite of information needed to evaluate ESG performance. But investors still need some independent sources of information beyond corporate disclosure to paint a more accurate picture of ESG risk and performance.

All this means that MSCI's ESG rankings, as rigorous as they are, still depend heavily upon subjectivity. Wherever standardized reporting data remains insufficiently robust for the comparisons they are striving to make, MSCI uses proprietary data sources and qualitative judgments to close the gap.

The problem is that highly subjective, qualitative considerations with a 50 percent weight leave lots of room for randomness.

Dane Christensen of the University of Oregon and George Serafeim and Anywhere Sikochi of Harvard Business School have looked closely at the quality and nature of corporate ESG disclosures and how they influence rankings by organizations like MSCI. A recent study of theirs found that the primary driver of differences in composite ESG ratings is the quantity and quality of corporate reporting, but not for the reasons you would expect.[5] According to their research, the greater the amount of information companies disclose, the more significant rating disagreements in materiality assessments and investment conclusions become—the exact opposite of what is being sought and what would be expected! This means that, to date at least, improved ESG disclosures are leading to less clarity about relative risks and valuations, not more. Christensen, Serafeim, and Sikochi also discovered more significant disclosures, and the rating disagreements they engendered had an essential and direct impact on stock and bond price movements. In sum, they found increased ESG disclosures made corporate valuations far more *volatile*, an outcome which may impact their access to reliable public financing. Again, this is not what one would expect: greater ESG disclosures to date have led to *more significant ESG rating disagreements*. Apparently, no good deed goes unpunished, even if you are a corporation. More significant disclosures often end up penalizing the disclosers, at least in the short run. One presumes that—over time, as other companies disclose more—these discrepancies should abate.

Perhaps this also helps explain why MSCI uses considerable qualitative judgment to aggregate and assign relative ESG grades. MSCI ultimately settled upon a rules-based methodology and alphanumeric scoring, ranking corporations by industry into three broad qualitative categories: Leader, Average, and Laggard. Leaders are rated AA or higher, Laggards are rated single B or lower, and Average ratings are those between AA and B, as shown in figure 7.2.

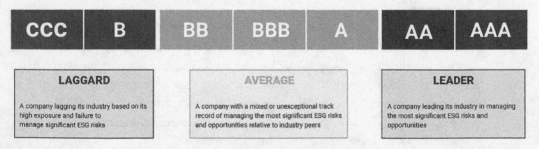

| CCC | B | BB | BBB | A | AA | AAA |

LAGGARD	AVERAGE	LEADER
A company lagging its industry based on its high exposure and failure to manage significant ESG risks	A company with a mixed or unexceptional track record of managing the most significant ESG risks and opportunities relative to industry peers	A company leading its industry in managing the most significant ESG risks and opportunities

FIGURE 7.2. MSCI's primary ESG scoring outputs.

As ESG ranking systems go, MSCI's deserves significant credit for measuring material risks as systematically and accurately as available data allows.
They also deserve accolades for creating a ranking methodology that is both
robust and intuitive: after all, what investor or financial analyst doesn't
understand their ABCs? When you drill down a bit deeper, however, one
fairly wonders how the rest of their subjective valuation process works. One
also reasonably asks whether an alphanumeric system that depends on large
amounts of proprietary data and qualitative judgments is sufficiently robust
to drive trillions of dollars in individual security selections and asset allocations on its own.

Take again, as an example, MSCI's mid-2021 rankings for the two fierc-
est soft drinks competitors, Coke and Pepsi. If you were to look at their
detailed, side-by-side ESG scores, Pepsi would appear to beat Coke in relative materiality risks. After all, Pepsi outscored Coke in four categories: nutrition and health, corporate governance, water stress, and corporate behavior.
They underperformed in only two: product safety and quality and health
and safety of workers. If two points were awarded for every ESG category
in which the firm is a leader, one point awarded for being average, and one
point subtracted for being a laggard, Pepsi would easily beat Coke, 12 to 9.
After MSCI's qualitative assessments kicked in, however, Coke was somehow deemed superior. When this chapter was first written, Coke ultimately
earned MSCI's coveted AAA rating, whereas PepsiCo merited only their AA
rating.[6] How this happened or precisely what tipped the scale is unclear. All
the numeric analytics that went into MSCI's system became subsidiary to the
qualitative judgments MSCI relied upon. The outcome is nevertheless significant for portfolio managers and analysts who count on MSCI for their relative
stock weightings and indices. Who, after all, would want to own a AA-rated
company in an industry in which a AAA-rated alternative is available?

Coca-Cola's stock has historically traded at a slight premium to Pepsi's
as measured by its price-to-earnings (PE) ratio. Stocks with higher PE
ratios generally reflect the market's expectation of superior future earnings
prospects. In their original 2021 rankings, MSCI apparently believed Coke
deserved such a premium, too. In fact, their proprietary model may well be
one of the factors that contributed to Coke's historically superior valuation
versus Pepsi.

Creating a single, transparent, actionable methodology for assigning
meaningful ESG scores is especially challenging given data is not fungible
between industries or between companies in the same industry. The absence
of consistent corporate reporting standards puts those paid to be rigorous

about their ESG investment process in a genuinely untenable position. Most data scientists say that if some data is good, more is even better. MSCI's rating process is good, arguably even the best—but 50 percent subjectivity still leaves considerable margin for disagreement. After all, MSCI's ESG ratings can never be better than the data and process upon which they ultimately depend, and we now appreciate ESG data reporting remains a work in progress. Given this, supplementing MSCI's approach with other methodologies, or even one's proprietary work, should make better ESG valuation decisions. MSCI is not prescient—no one is—nor does MSCI claim to have unlimited resources. Other investment and accounting firms may have some comparative advantages in terms of corporate access or analytics. Investors looking for the fullest picture should consider those, too.

One of the earliest efforts to categorize and measure material ESG risks systematically was undertaken by Sustainalytics. Michael Jantzi, the founder of Sustainalytics, first started publishing research on responsible investing in 1992 and has remained a market leader on sustainable investing ever since. In 2010, Jantzi was given a lifetime achievement award by the Responsible Investment Association. Over the years, Sustainalytics has been recognized as a leading, independent responsible investment research firm in multiple industry surveys.

The approach Sustainalytics takes to measure and assign ESG scores to corporations is fundamentally different from MSCI's (figure 7.3). Sustainalytics begins by identifying every material ESG risk each industry and company is exposed to and then segments those risks into two categories: those that can be mitigated and those that cannot. It seems a promising idea; after all, corporations, like people, should be judged by what they can control, not by what they can't. Take gold mining as an example. Gold mining firms have more material environmental exposures than software or specialty retail firms. Given the nature of their work, not all of those risks can be mitigated; you must go underground to look for gold, after all. Sustainalytics rigorously measures how effectively each company manages the material risks that are within their control. In the gold mining example, even though it is an environmentally challenged industry, some mining companies do a much better job than others at, say, minimizing caustic chemical use or promoting workplace safety. Sustainalytics gives companies that address the risks they must incur mindfully preferential intra-industry scores. They then reward the better actors.

The advantages of a numeric ranking system over an alphabetical grading system are also evident. With numbers, one company or industry can be compared quite tangibly to another, and the quantitative differences between

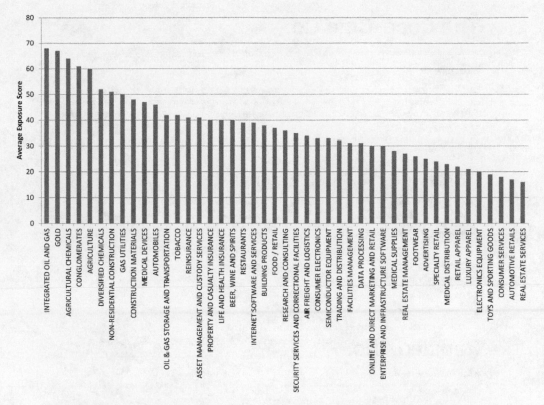

FIGURE 7.3. The Sustainalytics numeric ESG ranking system spans
forty-five industries.
Data: with permission from Sustainalytics.

companies and industries can be more immediately grasped. One can also
better measure a group of companies in a portfolio with different numeric
metrics, using average or median calculations. For example, it's easier to
assess the relative risk of 80 versus 20 on a 100-point scale than it is to rank
an A-rated versus a B-rated company. The former is entirely quantitative,
whereas the latter is qualitative and more subjective.

Still, suppose we return to our example of Coke versus Pepsi. This case
helps demonstrate the challenges of using only one ESG ranking system to
make large tactical weightings or tilts between firms in or across industries.
It also helps underscore the findings of Christensen, Serafeim, and Sikochi.
Unlike MSCI, which has marginally preferred Coke to Pepsi over time, it
turns out Sustainalytics prefers Pepsi to Coke—not by a little but by a lot
(figure 7.4). In their 2021 rankings, Sustainalytics does not even have the

The Coca-Cola Co.

Industry Group: **Food Products**   Country: **United States**

Identifier: **NYS:KO**

Coca-Cola is the largest nonalcoholic beverage entity in the world, owning and marketing some of the leading carbonated beverage brands, such as Coke, Fanta, and Sprite, as well as nonsparkling brands, such as Minute Maid, Georgia Coffee, Costa, and Glaceau. Operationally, the firm focuses its manufacturing efforts early in the supply chai...
+ Show More

Full time employees: **80,300**

ESG Risk Rating COMPREHENSIVE

22.5 Medium Risk

Negligible	Low	Medium	High	Severe
0-10	10-20	20-30	30-40	40+

Ranking

Industry Group (1st = lowest risk)
Food Products   **43** out of 598

Universe
Global Universe   **4697** out of 15096

Last Update: **Oct 5, 2021**

Pepsico, Inc.

Industry Group: **Food Products**   Country: **United States**

Identifier: **NAS:PEP**

PepsiCo is one of the largest food and beverage companies globally. It makes, markets, and sells a slew of brands across the beverage and snack categories, including Pepsi, Mountain Dew, Gatorade, Doritos, Lays, and Ruffles. The firm uses a largely integrated go-to-market model, though it does leverage third-party bottlers, contract manufacturer...
+ Show More

Full time employees: **291,000**

ESG Risk Rating COMPREHENSIVE

16.0 Low Risk

Negligible	Low	Medium	High	Severe
0-10	10-20	20-30	30-40	40+

Ranking

Industry Group (1st = lowest risk)
Food Products   **5** out of 598

Universe
Global Universe   **1638** out of 15096

Last Update: **Oct 7, 2021**

FIGURE 7.4. Hmm. Sustainalytics says Pepsi's better than Coke—a lot better.
With permission from Sustainalytics.

Coca-Cola Company in the top 10 percent in their field, whereas PepsiCo ranks in the top 1 percent. Overall, Coke merits a medium-risk grade from Sustainalytics, while Pepsi is a low risk. In fact, of the fourteen thousand companies Sustainalytics systematically graded, 3,615 companies separate the two. Thus, Sustainalytics effectively recommends you overweight Pepsi in your portfolio until it carries a premium versus Coke—the opposite of the valuation suggested by MSCI.

This simple but revealing example of how two sophisticated purveyors of ESG scores come to fundamentally different conclusions over two highly transparent and well-researched companies—Coke and Pepsi—offers some idea of how difficult assigning reliable ESG ratings can be while disclosed ESG data is still evolving. When copious quantities of data and evidence are available, two market-leading ESG-rating agencies can and do come to opposite conclusions. What Christensen, Serafeim, and Sikochi found at a macro level, we've just discovered on a micro level. One wonders what this must mean for companies and industries for which accurate data is harder to come by, and where subjective judgment is far more determinant in final scores, by necessity.

MSCI and Sustainalytics are not the only rating agencies battling it out to provide superior ESG insights, of course. In addition to MSCI and Sustainalytics, there are more than a dozen additional, serious vendors of primary ESG data and rankings. Some—like Refinitiv, Fitch Solutions, S&P Global, and the Sustainability Accounting Standards Board (SASB)—provide broad measurements that span multiple material risk categories across all three environmental, social, and governance spectrums. Others—like Vivid Economics, GlobalData, the Climate Bonds Initiative, and the Carbon Disclosure Project—are far more focused on specific risk classes, like climate. One example of the latter—Trucost—is now part of a more prominent index research firm, S&P Global. In practice, Trucost has intentionally focused its efforts on carbon-related concerns, including market-leading work on physical risks and natural resource constraints. Because of Trucost's superior analytics, many investment firms rely more heavily upon them than MSCI or Sustainalytics for climate insights.

My Coke-versus-Pepsi illustration is not unique. In addition to the work of Christensen, Serafeim, and Sikochi, other rigorous studies have looked at how and why ESG rating providers differ across multiple industries. One by Florian Berg of MIT, Julian Kölbel of the University of Zurich, and Roberto Rigobon also of MIT found the correlation between ESG scores of different providers to be only 0.54. This means ESG ratings by different purveyors are widely divergent. Correlations are even lower when considering individual "E," "S," and "G" pillars.[7] These results stand in marked contrast to those of credit rating agencies. Credit rating correlation between Moody's and S&P is a remarkable 0.99. In short, contemporary ESG analysis is clearly a work in progress. Methodologies and outcomes are widely divergent and seem likely to remain so for quite some time. ESG scoring involves some part art, some part science.

In another definitive report canvassing ESG investing practices, progress, and challenges, the OECD provides a compelling visual depiction of how inconsistent ESG rating methodologies produce incongruous results.[8] As you can see in figure 7.5, the ratings of ten corporations from eight industries scored by five ESG rating providers show little to no consistency, especially

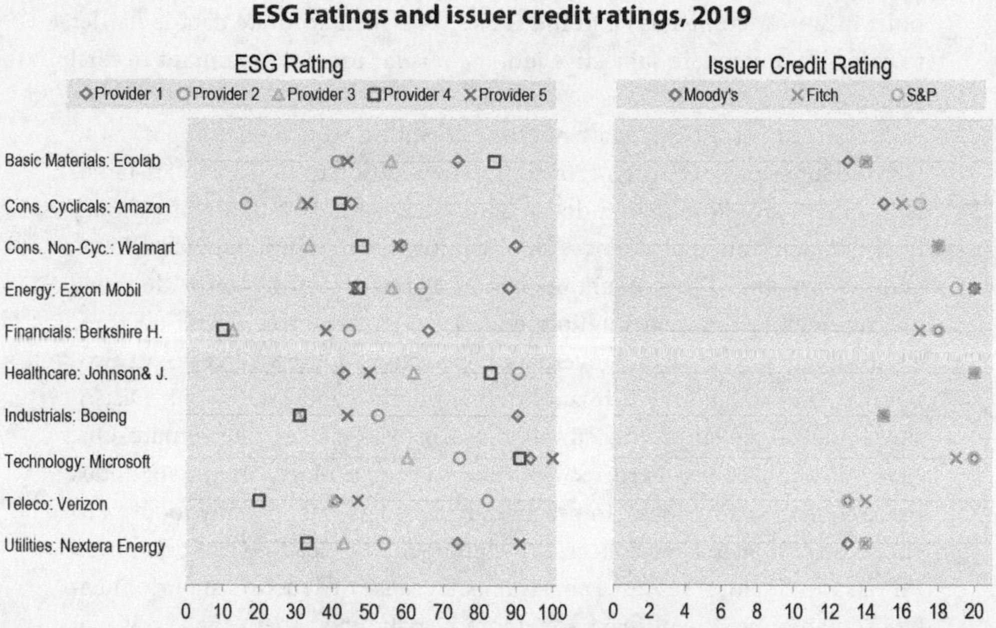

ESG ratings and issuer credit ratings, 2019

Note: Sample of public companies selected by largest market capitalisation as to represent different industries in the United States. The issuer credit ratings are transformed using a projection to the scale from 0 to 20, where 0 represents the lowest rating (C/D) and 20 the highest rating (Aaa/AAA).
Source: Refinitiv, Bloomberg, MSCI, Yahoo finance, Moody's, Fitch, S&P; OECD calculations

FIGURE 7.5. Something's not right: ESG ratings are idiosyncratic, whereas credit ratings are close.
Boffo, R., and R. Patalano (2020), "ESG Investing: Practices, Progress and Challenges," OECD Paris, www.oecd .org/finance/ESG-Investing-Practices-Progress-and-Challenges.pdf.

relative to their corresponding credit ratings shown on the right side. There really should be no surprise with these results, though. Whenever data and methodologies are broadly different, outputs will be too.

If, as it very much seems, better data would lead to better indices and more informed decision-making, it follows that more effort should be expended to improve data. As mentioned earlier, an essential aspect of creating better data is more precise standardization. Efforts to standardize data collection from corporations have long been an objective of both private sector and public sector bodies working in collaboration. For example, the Financial Accounting Standards Board (FASB) is an independent, not-for-profit organization most widely recognized for its rigor in promoting financial reporting practices for both public and private companies. FASB is also recognized by the U.S. Securities and Exchange Commission as the designated accounting standard setter for public companies. Many other accounting organizations also recognize FASB standards as the most authoritative, including the American Institute of Certified Public Accountants and most state Boards of Accountancy. In short, FASB is the financial accounting gold standard. This means that failing to abide by FASB standards is risky and viewed as a red flag by most investment analysts and auditors. FASB also has a sister organization, the Governmental Accounting Standards Board (GASB). GASB sets reporting standards for state and local government organizations. FASB and GASB broadly promote the Generally Accepted Accounting Principles (GAAP), the framework endorsed by the U.S. Securities and Exchange Commission. GAAP standards often differ considerably from the global standard known as the International Financial Reporting Standards (IFRS).

By now I fully expect you've stopped paying any real attention to what you've been reading. Regrettably, deep dives must go deep. Let me bring you back to my core argument by simply highlighting that *there are no broadly agreed accounting standards on emissions, worker safety, employee pay or satisfaction, diversity, or any other primary environmental or social concerns.* In other words, virtually none of the prior discussions on what MSCI, Sustainalytics, or any other serious ESG data providers are up to have yet been incorporated into the most recognized accounting standards bodies like FASB or GASB. It is then perhaps no surprise to learn that a new private, nonprofit initiative has sprung up to improve accounting standards relating to sustainability. Aptly named the Sustainability Accounting Standards Board, its materiality map looks much how you would expect: detailed and complicated. Investors focused on material ESG risks clearly need compilers of fungible data like SASB to succeed. Over time—years, certainly not months—I hope they will.

SASB was formed in 2011 with much the same structure and oversight as FASB. SASB ultimately hopes to surface all information about sustainability that is financially material—that is, of potential pertinence to a company's earnings and ultimate valuations (figure 7.6). Remember, they are not trying to determine whether a company is good or just or balanced or ethical. Instead, they seek industry-specific information that is useful for making decisions and evidence based, all sourced in a highly standardized form that enables investors to make meaningful judgments about relative *financial* risks generated by environmental, social, and governance concerns. It's a laudable initiative. It's also sorely needed. This said, were SASB a baseball game, it's probably only in its first or second inning. Of the more than fifty thousand publicly traded companies globally, only about one thousand have regularly referenced SASB in their annual disclosures (figure 7.7), and a mere five hundred or so have fully complied. Understandably, most asset managers and many regulators are urging faster SASB adoption— remember, for investors, more transparency is always good. But some of those financial firms have yet to adopt SASB protocols themselves. We need everyone to report more data in more conformable manners so true comparisons can be drawn.

A final, significant effort seeking to generate standardized information relating to climate risk is the Task Force on Climate-Related Financial Disclosures (TCFD). Chaired by the former mayor of New York and billionaire philanthropist Michael Bloomberg, the TCFD was effectively commissioned by the Financial Stability Board (FSB). FSB aims to promote international financial stability by persuading national financial authorities and international standard-setting bodies to collaborate on shared priorities. FSB established the TCFD in direct response to a request by the G20 to better assess how all issues relating to climate change pose risks to global financial stability. The G20 is the closest thing humanity has to a global decision-making body. If there is ever to be successful global cooperation on climate, regulation, and accounting standards in the modern world, the G20 will need to play a determinant role. To date, however, they have failed to do so.

The most important contributions of the TCFD so far have been to highlight recommended standards for reporting financial risks relating to climate, including climate stress test scenarios. Theirs is a serious effort,

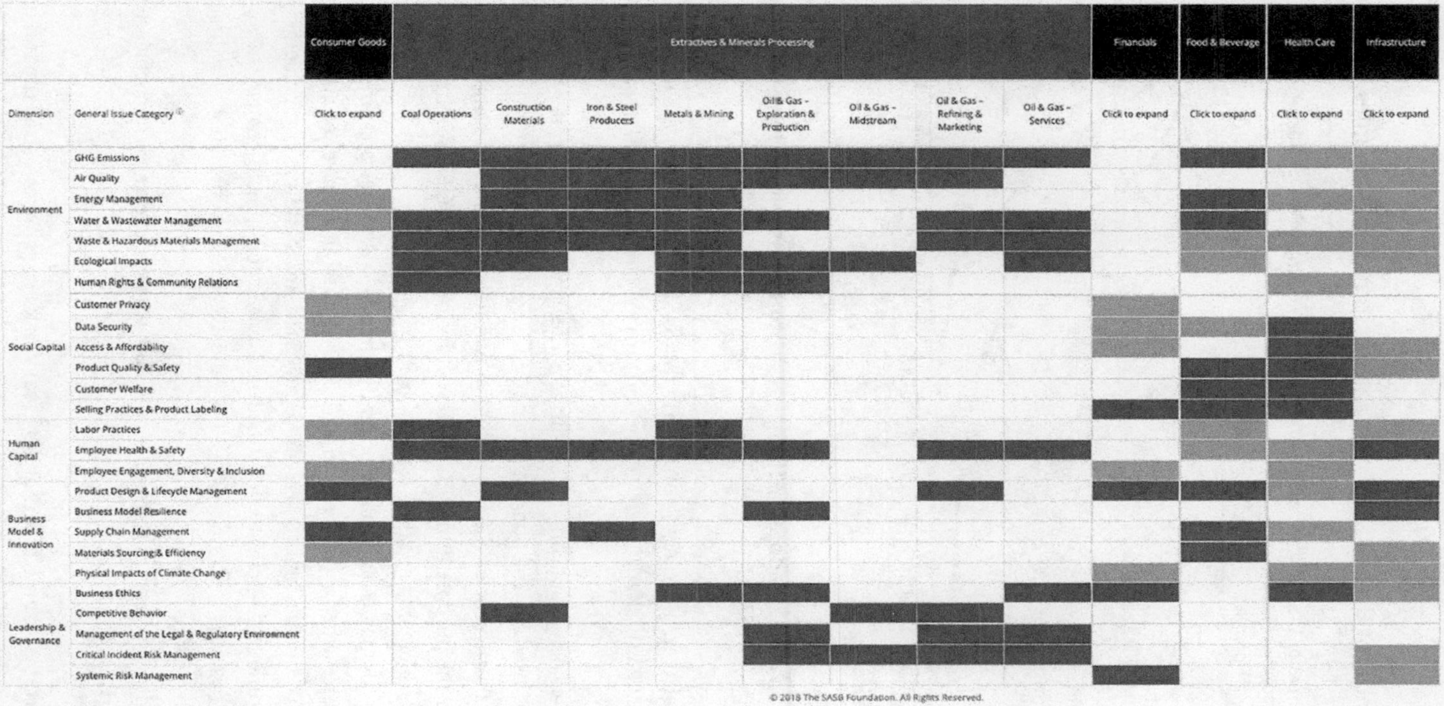

FIGURE 7.6. SASB's evolving Materiality Map.

Number of Unique Reporting and Referencing Companies, by Year

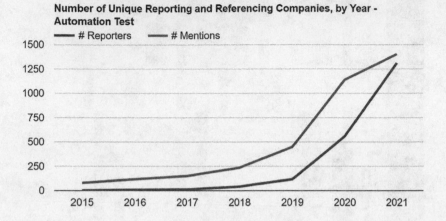

Number of Unique Reporting and Referencing Companies, by Year - Automation Test

Reporting companies account for a subset of referencing companies.

FIGURE 7.7. SASB adoption is important—but has a long, long way to go.
Reprinted with permission from Value Reporting Foundation. All rights reserved.

but it is also in its early innings. In the TCFD's 2021 annual update, five milestones were cited:

1) Forty-two percent of companies with a market capitalization greater than $10 billion disclosed information in line with individual TCFD recommendations.
2) Nearly 60 percent of the world's one hundred largest public companies either support the purpose of the TCFD, report in line with TCFD recommendations, or both. (Note: Mainly, they support them.)
3) Energy companies and materials and buildings companies are now leading on disclosure, with an average level of TCFD-aligned disclosures of 40 percent for energy companies and 30 percent for materials and buildings companies.
4) Expert users of disclosure identified the impact of climate change on a company's business and strategy as the "most useful" information for financial decision-making.
5) Asset managers and asset owners reporting to their clients and beneficiaries on TCFD risks remain deficient.

In short, SASB and TCFD are critical works in progress. And the net effect of all these efforts—from MSCI and Sustainalytics to SASB and TCFD—are best summarized by one word: *incomplete*.

So what are we to make of all these efforts from an investment standpoint? Given data sets are incomplete and methodologies so divergent, how reliant should we become on any single ESG algorithm, process, or score?

Then SEC chairman, Jay Clayton, offered his own views on ESG analytics in 2020. It wasn't encouraging. He said he had "not seen circumstances where combining an analysis of E, S, and G together, across a broad range of companies . . . with a single rating or score, would facilitate meaningful investment analysis that was not significantly overinclusive and imprecise."[9] Like me earlier, Clayton seems to have suggested an ESG divorce may be better for all concerned. If the former head of the SEC came to this conclusion—saying in effect that composite ESG scoring systems are by themselves ill-equipped for conducting meaningful investment decisions—perhaps we should all be more circumspect about what actionable information they provide.

To be clear, ESG risks are real. Multiple environmental, social, and governance concerns are undoubtedly financially material. These risks are sure to drive asset prices in meaningful ways for the foreseeable future. Time will bring improvements in ESG reporting standards and more robust applications in investment products and practices. We should all welcome these improvements. ESG data collection is at an early stage, and its further maturation is important. The realistic time horizon for realizing all needed improvements will be measured in years, though, not weeks or months. What will occur between now and then is also quite clear: trillions of dollars will remain in motion as multiple proponents claim to have unique processes, insights, and proprietary investment techniques that can generate superior results. Many will claim they have somehow found the best ESG index or metrics for measuring risks and achieving superior returns. Some of these claims will be right, of course. But many others will be wrong. In a highly competitive industry, it is simply not possible for everyone to be the best at everything. What we can reasonably predict is that the top 10 percent will be top decile, and one of four will be top quartile. We can also predict with certainty that there will be bottom deciles and bottom quartiles. This is how investment competition works. Given how much is at stake, you can bet considerable efforts will be expended to gain an edge over all other ESG analytics and product purveyors.

The technical process of ESG index creation, along with emerging efforts seeking to measure how values can be predictable indicators of stock valuations, will also continue to evolve and, in certain cases, generate significant outperformance. Understanding how these indices work, as well as emerging ESG valuation techniques, will bring us to the cusp of understanding what

I refer to as the raging "ESG arms race." As you might expect, asset management firms are now locked in a battle to win a disproportionate share of the trillions of dollars that are being shifted from traditional investment products to those that carry an increasingly coveted sobriquet, and a highly sought-after investment outcome: *sustainable*.

But before we go there, it's essential to return to our core thesis and the ultimate purpose of this book. We began by agreeing that the globe urgently needs more inclusive, widespread, and sustainable prosperity. We subsequently committed ourselves to ferreting out the best financial tools and strategies to promote such inclusive, sustainable growth. Obviously, many ESG strategies lay claim to this mantle. Multiple ESG proponents explicitly or implicitly promise double-bottom-line outcomes—that is, better environmental and social outcomes *and* better returns. If ESG is going to work, great: we're on the right path. If it is not, however, something else will be needed. Right now, the jury is still out.

But given all we've just learned, how can any investment strategy be sustainable if climate change continues unabated? If environmental and social upheaval lies ahead, is there any asset class or method for us to invest our money that could remain beyond disaster's reach? How can we ever know which investment strategy is best without first assessing which companies have successfully insulated themselves from mounting ESG concerns?

It turns out there are a number of serious efforts underway to answer all these questions. A group of highly sophisticated financial alchemists believe they can bundle efficient, effective solutions. They are also broadly underappreciated and operate out of public view: those who design and maintain stock and bond indices.

Chapter Eight

A FEW WORDS ABOUT INDICES

Don't look for the needle in the haystack; just buy the haystack!
JOHN C. BOGLE

I used to think, if there is reincarnation, I want to come back as the
president or the pope or a .400 baseball hitter. Now I just want to
come back as the bond market. You can intimidate everyone.
JAMES CARVILLE

How well do you remember the early 1990s? Madonna and Michael Jackson
were in full stride. The best-selling car was the Honda Accord, registering
twenty-one miles to the gallon. Ten-year Treasury yields were north of 8
percent. When Bill Clinton came into office, his chief political adviser, James
Carville, was shocked to learn how constrained his boss's fiscal policy options
were. If yawning budget deficits were not closed and federal spending not
paid for with politically risky tax hikes, he was told, interest rates would
spike higher, derailing economic growth. It's why Carville said he wanted to
be reincarnated as the bond market, rather than as the president or pope.

But James Carville's historic fantasy is sorely out of date. Now that central
banks resort to quantitative easing techniques whenever they dislike the level
or shape of the yield curve, today's highest financial power no longer resides
in the bond market; it has shifted to the stock market. More specifically, it
has shifted to the index makers who define how we think about the stock and
bond markets. Their computations and compilations literally drive trillions
of dollars of investment decisions daily. Index providers ultimately define
success and failure by defining how financial performance gets tracked and
measured. If we are to continue our investigation of how finance can help
forge the good society, we'll need to understand how stock and bond indices
are made and updated. Fortunately, it isn't all that hard to do. It all began
with a simple, inspired idea from Charles Dow.

In the late eighteenth century, most of the traders who gathered in or outside the Tontine Coffee House on the corner of Wall and Water Streets in Lower Manhattan relied upon the credibility of their confreres, hearsay, and gossip as they bought and sold shares, debt certificates, and contracts from and to one another. Since the historic Buttonwood Agreement of 1792 limited the number of qualified financial intermediaries, Tontine is where New York's accredited merchants and financiers met informally, usually around 10 a.m., to barter scraps of paper and partnerships for cash and promises (figure 8.1). It was a relaxed affair at the beginning. Over time, as volumes escalated, they relocated to 11 Wall Street, the current home of the New York Stock Exchange.

In 1880, with his partner, Edward Jones—a recent dropout of Brown University—Dow began publishing the *Customer's Afternoon Newsletter*. This concise, entirely factual two-page summary sated a growing desire for something that had never been published before: timely, accurate financial data. In short order, their daily readership jumped to more than one thousand, and their reporting staff expanded to eighty-nine. Their vision and focused execution culminated in the first edition of the *Wall Street Journal* on July 8, 1889. Selling for two cents a copy or $5 for an annual subscription, its motto, "The truth in its proper use," described both its ambition and constraints. "Truth" precluded editorials or opinions of any other kind for that matter. Facts were prized, as was exact data. Fictions were despised.[1]

One of Dow's signature innovations was the numeric reporting of where shares of the most important companies of the day had traded. Every morning, beginning in 1884, Dow's reporters ferreted out the prices at which nine of the largest railroads, the largest steamship company, and the telegraph operator Western Union had most recently exchanged hands. They summed these prices, divided the total by 11—and voilà, the first stock index, was born!

It is easy to appreciate how the regularity and transparency of this new Dow Jones stock index conveyed something unique and valuable: a quantitative summary of the broad sentiment of the most informed merchants and traders of the day. As the index price trended higher or lower, one could readily infer how commercial confidence was waxing or waning. Over time, one could also see how fortunes were being made, lost, and made all over again. For the first time, one had some conception of how "the market" overall was performing. Before there was a Dow Jones index, there was no "market" per se.

FIGURE 8.1. Trading stocks under the buttonwood tree, circa 1800.
Bettmann, Getty Images.

Fast forward 150 years, and the calculation and publication of stock, bond, and commodity indices have become a massively complex, multi-trillion-dollar business. When Dow's *Customer's Afternoon Newsletter* commenced publication, there were no more than a few dozen U.S. companies of significance and a few thousand citizens sufficiently well heeled to care much about them. Today, there are about fifty thousand publicly listed stocks around the globe and tens of millions of institutional investors and individuals who care deeply about their price movements. As we know too well, the "stock market" has also now become so crucial that most daily newspapers and evening news broadcasts feel compelled to report how much it has fluctuated over the past twenty-four hours. They can do so only by quoting some reference index. Millions of computer terminals and mobile devices also stream the stock market's every movement perpetually, in real time. At any given moment, these indices allow one to fathom how the market is doing.[2] There are also multiple trillions' worth of publicly traded bonds that have been bunched together and quoted as bond indices. All this helps explain why, surprisingly perhaps, there aren't just thousands, but rather *millions*, of indices today. A total of 160,000 of these indices belong to Dow Jones alone.

The idea of millions of indices should sound utterly crazy to you. Why, every sensible individual should ask, would the world need millions of indices when there are only fifty thousand or so stocks? Isn't the former merely a derivative of the latter? The answer is not that complicated, however. Index creators and financial news publishers assemble and distribute multitudes of index values and volumes for the very same reason Charles Dow first started to do it in 1884. First, many folks are keen to have transparent, quantitative indications of baskets of asset price movements in real time and over time. For example, portfolio managers need indices to portray the investment outcomes they are trying to achieve accurately. Second, and perhaps most importantly, multiple index users are willing to pay providers to create indices to track the outcomes they most want to achieve.

Several vital index innovations occurred over the last century and a half to enumerate investor sentiments and asset price movements more accurately and actionably. For example, Dow's first index weighted the *prices* of the Erie and the Baltimore and Ohio Railroads equally, even though the B&O had much larger users and revenues. And while Western Union was the most crucial telegraph company of its day, it made a rather odd bedfellow with nine railroads and a shipping company. For an index to convey what was genuinely transpiring in, say, the transportation sector, it would make more sense to weight companies proportionately—that is, relative to their total income or enterprise value. It would also make sense to exclude Western Union from the calculation, concentrating exclusively on shipping and rail rather than diluting them with a communications company.

Unsurprisingly, over time, this is exactly what happened. As markets evolved, indices were developed to portray industry specializations, like industrials, financials, or technology. National identifications also emerged; hence, in the United Kingdom, there is the Financial Times Stock Exchange (FTSE) 100 Index and in Hong Kong, the Hang Seng Index. Equity and bond indices were further segmented into so-called developed, developing, and frontier markets. Today, more than six dozen countries have multiple stock and bond indices, and those indices have been further demarcated into industry groupings or segmented by other corporate behavioral character- istics like "growth" or "value." The calculation of indices also evolved from price weightings—Dow's preferred, though rather misleading, approach—to market-cap weightings or equal-price weightings. The focus on market weighting rather than price weighting meant stock market indices became more representative of actual economic activity. The most significant, impactful companies were given bigger, more impactful weights, directly correlating to their economic impact. While different indices portray differ- ent insights, they all provide real-time, transparent valuations on the under- lying assets upon which they are based.

Tracking the evolution of certain equity indices is a lot like reading economic history. In the latter half of the nineteenth century, the United States saw a historic expansion of industrial activity, facilitated in no small part by rapidly expanding transcontinental rail lines. Financing these new lines and combining existing routes usually took place at the New York Stock Exchange. Between 1860 and 1890, the U.S. population doubled, from thirty-one million to sixty-two million. This meant demand for commodi- ties and finished products grew exponentially through the late 1800s. To better distinguish rail and shipping from other industry trends, Charles Dow segmented his stock indices in 1896 between industrial and trans- portation companies, including twelve firms in each index. Dow's origi- nal twelve industrial firms included the United States Leather Company, the United States Rubber Company, the American Tobacco Company, and American Cotton Oil Company. These names probably seem unfamiliar because none of their descendants survive today. Only one company from

Dow's original industrial twelve—General Electric—lasted more than fifty years.[3] The Dow Jones Industrial Average (DJIA) expanded to thirty companies in 1928 and has remained at thirty ever since. Why thirty, you may ask? Why not thirty-five, or perhaps an even one hundred? No reason, really; it's just a number the index creators at the *Wall Street Journal* seem to have grown fond of.

You'll recall from chapter 2 the composition of the DJIA 30 in 1970 when Milton Friedman presciently fretted about corporate America's declining profitability and diminishing economic prospects. Obviously, the index represented a very different economy back then. About half of the thirty corporations in the 1970 DJIA mined minerals, smelted metals, and produced chemicals, including three that exclusively drilled for and refined oil. Six produced autos, auto parts, aircraft, or other durable goods. Four others— General Mills, Sears, Woolworth, and General Foods—primarily catered to U.S. consumer needs. Only one stock of the DJIA 30 of 1970, when Professor Friedman urged businesses to focus more on their profits, remains today: Procter & Gamble.

This historical depiction of the Dow Jones stock indices in constant flux with no apparent methodology behind the decisions helps debunk what I consider to be one of the most misleading and dangerous financial myths commonly espoused today, namely that index investing is somehow *passive*. As you can now appreciate, there is nothing passive about the DJIA 30, or any of the other major equity indices for that matter. Passivity means nothing changes. Passivity is static, suggesting the driver is asleep at the wheel. Obviously, this is not the case. The firms that compose stock market indices are habitually added and replaced. Their weightings are also ever-evolving. To say index investing is passive is like saying my wife is married to the same person she walked down the aisle with in 1989. To say index investing is passive is like saying every NBA championship team has had equal talents. To say index investing is passive is like saying every eighteen-hole golf course is alike. Index investing isn't passive; it's indexed.

One reason it is so hard to "beat the market" relates to the ways and frequency with which different stock indices are created, refreshed, and renewed. Many folks presume they can be more intelligent than a broad index because they have some unique gift. Well, those folks are almost always wrong. The more stocks in an index, the more likely it replicates the broad economic area it intends to represent. Probability and the law of averages tilt the table away from active stock pickers over more extended periods. In fact, fewer than 15 percent of active stock managers—just one of seven—have beaten a broad index over any five-year period.

Given that there are millions of indices to choose from, the index selection one ultimately makes is highly consequential. It's a lot like choosing a car. Drivers know the differences between a Toyota and a Tesla or between a Buick and a Bugatti before buying one because they research them intensively. Likewise, investors should be equally discerning when choosing a stock or bond market index to track.

Today, more than 120 institutions create and maintain various investment indices, from niche players like WisdomTree and the University of Chicago Booth School of Business to behemoths like MSCI. Four dominate the field: S&P Dow Jones, MSCI, FTSE Russell, and Bloomberg. Each has its own methodology, and each of those methodologies has distinct advantages and disadvantages. Some, like Russell, use objective, transparent algorithms to select and remove companies. Others rely upon subjective judgment. For example, all the companies that compose the S&P 500—the most widely tracked stock market index globally—are selected by a committee, just like the DJIA. Frankly, the process is not dissimilar to how popes get elected, or Oscars get awarded. In the case of the DJIA, the editors of the *Wall Street Journal*—being something akin to a financial version of the College of Cardinals or the Academy of Motion Picture Arts and Sciences—assemble, debate, and ultimately decide. Once they agree upon reconstitution, the equivalent of white smoke comes out of their chimney. The selection committees obviously reference some data—things like market capitalization or monthly trading volumes. But even these may not be enough to be selected. Some subjectivity remains. That's why, late in 2020, investors waited with bated breath to learn if Tesla—the electric vehicle darling, already then with a market capitalization of more than 99.9 percent of all other listed companies in the world—would somehow be added to the S&P 500. Tesla's apparent sticking point for election by the S&P committee had been the irregularity of its earnings. Making money is something the S&P team takes pretty seriously. They were impressed by Tesla's stock performance, but, like many, they wondered if its business model would ever generate actual profits. Elon Musk had intentionally made Tesla into a golden apparition, with profit promises always in the distant future. Well, Tesla's white smoke appeared on November 30, 2020—and twenty-one days later, every stock in the S&P 500 had to be readjusted to make room for their newly admitted, by then $600 billion baby brother to join.

The S&P 500 is not passive. It adjusts to economic realities as quickly as its compilers can ferret them out. And figuring out economic realities is something S&P's index providers work very hard at.

Unlike the S&P 500 or the DJIA, Russell indices are constructed with a more objective methodology. They also involve many more companies—three thousand in the United States alone. Because they include many more companies, Russell indices have justifiably become the industry standard for smaller and midsized firms. Requirements include where a corporation is headquartered and listed, its minimum market capitalization ($30 million at this writing), and whether 5 percent or more of its shares are freely traded. Russell indices are also reconstituted according to a set schedule: once a year, on the last Friday in June. As a result, trading volumes on reconstitution days spike massively, sometimes by tens of trillions of dollars. Russell's U.S. indices capture approximately 99 percent of the U.S. equity market and nearly 100 percent of its investable universe.

As you might imagine, adding companies to and deleting them from an index has significant implications. Ascension conveys growing importance; inclusion, ongoing legitimacy; and demotion, a harbinger of darker days to come. Ascension also means automatic inflows from those who use that index as their benchmark, whereas demotion means the converse. Thus, in 1999, when declining Chevron, Goodyear Tire, Sears, and Union Carbide were dropped from the DJIA, and ascending Home Depot, Intel, Microsoft, and SBC Communications were added to fill their places, lots of buying and selling occurred axiomatically. This is what I meant by James Carville's outdated respect for the bond market. What you want to know today is, What stocks and bonds are going to be added and dropped from the most widely tracked indices? Given it's out with the old, in with the new, the consequences can be significant. Rejected stocks broadly decline as their execution dates approach; ascending names rally in anticipation of rising investor demand. Some popes and presidents may only dream of being so powerful!

All that said, index electors can and often do get plenty wrong. Consider the DJIA 30, again. IBM was first added in 1932, deleted in 1939, and then readded in 1979. For most of the forty-year tenure during which it was *not* included, IBM was an outstanding stock to own. Yet from 1999 to 2019, during which time it was included, IBM traded at or near the same price. This means the DJIA 30 compilers got IBM's share price movements back-to-front. And while IBM's stock has spent most of the past two decades effectively going nowhere, Amazon's has risen 4,450 percent. DJIA 30 adherents have essentially lost out on the e-commerce wave. IBM remains in the DJIA to this day. Amazon has never been added.

Over the DJIA's 125-year history, its composition has been reconfigured fifty-seven times. Sometimes those reconfigurations took place frequently—within a matter of months—but other times, no changes took place for years. In fact, over the thirty-seven years from 1939 to 1976, recomposition took place only once, with four companies being replaced. It's tough to imagine that the U.S. economy changed so little over those three and a half decades. But then, from just 2015 to 2020, the DJIA 30 was revised seven times. Again, it is somewhat like the pope; some names stay for decades, others for only a matter of months. Moreover, it turns out some DJIA 30 periods generate better relative outcomes, just like pontificates. John Paul II was undoubtedly among the best, and Stephen VI and Boniface VIII were arguably the worst, but clearly I digress.

The shifting fortunes of many U.S. corporations and industries are not the only things you can glean by tracing the history of stock indices. You can also see the vicissitudes of investor exuberance and pessimism. When the original Dow Jones index was introduced in 1884, it stood at 62.76. It reached a nineteenth-century peak of 78.38 in the summer of 1890 but fell to an all-time low of 28.48 during the Panic of 1896.[4] Over long periods of time, though, stock market performance has been an unrelenting tale of expanding fortune. The DJIA 30 exceeded thirty-six thousand eight hundred at the end of 2021, an annualized return of more than 8 percent since 1957, including dividends. Over that same period, the S&P 500 returned 9.25 percent per annum, again including dividends (figure 8.2), meaning you would have doubled your money every eight years. As long as economies continue to grow, broad stock market indices will almost certainly grow with them. Still, some indices will perform better than others. A lot of work continues on designing the best indices for tomorrow.[5]

While the DJIA remains one of the most widely quoted stock market indicators of all time, its relatively random and secretive method of composition as well as its rather poor correlation to the broader economy left significant opportunities for improvement to other index providers and statisticians. You'll recall Charles Dow's first index was weighted by stock *prices* rather than market capitalization or income. The DJIA 30 still is. Unsurprisingly, this methodology leads to some head-scratching outcomes. For example, the world's largest company by market capitalization—Apple—recently composed 2.5 percent of the total value of the DJIA, while Goldman Sachs's share

Dow vs. S&P 500 since 1957

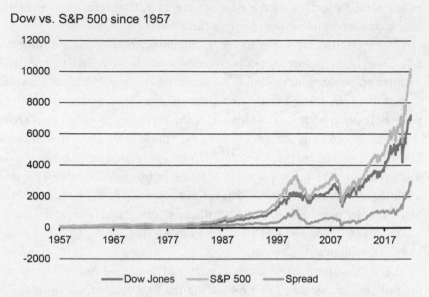

FIGURE 8.2. Which would you rather own: the DJIA 30 or the
S&P 500?
Figure by the author.

of the DJIA was about 6.5 percent. This means Goldman's stock price movement has more than double the impact on the DJIA's daily performance than
Apple, even though Apple's enterprise value is nearly fifteen times that of
Goldman. Why? Because Goldman's stock was trading around $400, whereas
Apple's was near $150. Never mind stock splits or the noncorrelation between
stock prices and corporate valuation. Never mind which matters more to
global GDP. For the DJIA, stock prices are all that matter.

Another innovation some index providers have used, in addition to
market-cap weightings, is equal weightings. In this case, whether there are
fifty or one hundred or three thousand stocks in an index, each is assigned an
equal component. And rather than creating index specializations by industry
or geography, some providers segment companies by the way investors treat
them or analysts evaluate them. For example, a great deal of academic work
in recent years has gone into classifying companies by the broad characteristics they manifest, such as quality of earnings, volatility, growth, value, or
momentum versus other stocks.[6]

Finally, and most pertinent for our current purposes, a growing number
of indices now segment companies by their ESG credentials, screening out

those presumed to have, say, large carbon footprints while elevating those with superior metrics for, say, their chosen form of governance or direct social contributions. As of December 2021, there were more than 3,500 separate ESG indices and millions of institutional and private investors who relied upon them for their investment allocations and financial returns. Unsurprisingly, MSCI has taken the leadership role in designing ESG indices and is responsible for about half of the field. Some are climate aligned, others faith based. In others, "sin industries" like controversial weapons and tobacco are omitted. MSCI also has an equity social impact fund that invests only in firms with female leadership and a bond index fund that includes only green bonds, so called since their proceeds must be used for an environmental purpose. In addition, many new ESG funds are being created based upon their alignment to the 2050 net-zero emissions target of the Paris Agreement. All of these indices have a specified purpose: to serve as a measuring stick versus some desired outcome. If a given stock or bond is in an ESG index and an asset manager is using that index, they need to own it; if they don't, they could underperform. You may ask, How have these new indices changed capital flows, impacted the environment, and improved social outcomes so far? It's genuinely impossible to say. Very little in the way of impact measurement has taken place. While nearly every corporation cares about their perception by core stakeholders—most especially their employees and shareholders—a smaller portion explicitly target better ESG scores. ESG capital flows are clearly impacting valuations, meanwhile—a topic we will return to in later chapters.

While Charles Dow envisioned the first index and Russell, MSCI, S&P, and Bloomberg ran further with his brilliant innovation, John Bogle undoubtedly deserves the most credit for putting indices to work. Bogle moved theory into practice. "My ideas are very simple," Bogle once said. "In investing, you get what you don't pay for. Costs matter. In time, intelligent investors will use low-cost index funds to build a diversified portfolio; they won't be foolish enough to think they can consistently outsmart the market."

There is a lot to unpack in this statement, but the key thing to understand is that Bogle was then and to this day remains right. As we've learned, stock indices are what allow us to even speak about "the market." On cost, Bogle refers to the fact that, since someone else is doing all the work to constitute an index, you don't need to replicate those efforts. Those who manage indexed investment strategies incur relatively negligible research expenses. Once you choose an index, you simply buy or sell what the provider includes or excludes. Passing these savings on to investors can have significant results over time. Active managers often charge as much as 1 to 2 percent of the capital they manage on an annual basis—meaning as much as 10 to 20 percent of your savings would be eaten up every decade, paid out in fees to an active manager. Fees are as low as three to five basis points with index managers, meaning no more than 0.03 to 0.05 percent is paid out to a manager over a decade. In a $10,000 stock portfolio managed over a decade, that would mean a fee savings differential of as much as $1,700. That's huge!

John Bogle's family endured personal hardships during the Great Depression, losing their home to foreclosure and John's father tragically to suicide. This may explain why Bogle committed himself so passionately to helping ordinary working families achieve greater financial security over time. His senior thesis envisioned how a low-cost investment company might enrich its customers rather than its managers. His life's work turned this thesis into one of the world's leading asset management firms.

Bogle was not the first person to launch an index fund, mind you—that privilege appears to belong to Richard Beach and Walton Dutcher of Florida in 1972. But Bogle was the first with the tenacity to stick with one until it became a success. When he launched his First Index Investment Trust in 1974, it was derided as "Bogle's Folly." No surprise, this: active managers had a lot to lose if he was ultimately successful. They attacked him in part out of fear. If Bogle was right, the vast majority of his critics would be put out of business. In fact, few "follies" have gone on to achieve similar acclaim.

Bogle based his first investment trust on the S&P 500. After garnering only a few million dollars in its first five years, folks finally began to notice he was on to something. Given that more than 85 percent of active asset managers fail to match the returns of the S&P 500 over any five years, time was all he needed to prove his point. Even more fail to outperform over ten years. Worse, most charge hefty fees for their studied underperformance. As only a handful of managers ultimately beat him, Bogle used his verifiable, expanding track record to prove his detractors were simply wrong. It turns out active managers trying to beat the market weren't better fiduciaries, producing better risk-adjusted returns at reasonable prices. Bogle was. His funds grew throughout the 1980s and 1990s and at an ever-accelerating pace following the global financial crisis. His first fund (since renamed the Vanguard 500 Index Fund) had a stunning market capitalization of more than $750 billion at the end of 2021. The Vanguard Group now manages more than $7 trillion overall. Index investing has in many ways become the asset management standard rather than the exception. In 1995, 5 percent of the U.S. equity mutual fund industry was formally indexed; by 2020, 50 percent was, a market share gain of ten times.

Before we turn our attention to methods that asset management firms use to generate superior risk-adjusted returns and the raging competition in the industry to win the "ESG arms race," it may also be helpful to dissect the business of asset management, at least a bit.

The asset management business is pretty simple when viewed through one lens. Success largely depends upon two variables: the volume of assets under management and the fees charged to manage them. Index investing does not require armies of highly paid analysts and star fund managers, so they can be offered to clients relatively cheaply—for single basis points rather than dozens or hundreds. However, to have a successful index asset management franchise, you need to have a lot of scale and volume. To have a successful *active* asset management business, in contrast, you need excellent performance and a competitive fee incentive structure. Diversity of product—between equities and debt and in domestic, developed, and developing markets—is also strategically helpful. If all you have is one cheap index product—say, a fund linked to the price of Bitcoin—and its underlying market goes sour, your company will go sour, too. Ditto if you have only one active product; a few bad years, and you'd probably have to pack up shop.

One relatively recent financial innovation that has allowed asset management companies to launch more targeted indexed products efficiently is exchange-traded funds (ETFs). The ETF market has grown massively in recent years, from a market cap of a few hundred billion after the global financial crisis to more than $8 trillion today. ETFs are an exceptionally efficient and cost-effective way to access precise index exposures. Every share of an ETF is a fractional ownership interest in a portfolio of securities, the vast bulk of which are linked to common indices or screened versions of standard indices. ETFs have characteristics that are a lot like stocks and mutual funds. Most are listed on stock exchanges and can be bought and sold multiple times a day, just like stocks. They are also often more tax efficient than a mutual fund because of the way they are managed. Traders use them to hedge their risks and access market liquidity. Today, ETFs are an essential component of the global capital markets and financial ecosystem.

According to data compiled by Bloomberg, ETFs garnered more than $1 trillion of inflows during 2021—more than any prior year on record. Over the same period, mutual funds added just $80 billion, but this was only because of bond funds; equity mutual funds declined. In 2020, ETFs attracted about $500 billion of new inflows, while mutual funds lost about that much. The switch from mutual funds to ETFs is mostly about cost—ETFs are much cheaper—but it's also about innovation. ETFs can be structured with exact exposures, screening out unwanted names or industries, targeting size or sector, or only including firms that exhibit particular characteristics like minimum volatility, momentum, or quality. This can also be done efficiently and cheaply. Precision, price, performance, and liquidity are compelling qualities for investment products to have. The ETF market has an especially bright future.

Given how hot ESG investing has become, it should also be no surprise that the biggest ETFs ever launched have had ESG themes. The largest ETF launch was by BlackRock, focusing on companies that are best positioned for the pressures of Paris alignment. Dubbed "low-carbon transition ready" and carrying the ticker symbols "LCTU" in the United States and "LCTD" for its global counterpart, these two funds raised nearly $2 billion at launch. The underlying indices they track are the Russell 1000 in the United States and the MSCI World Index, respectively. The funds' managers can over-weight stocks that are considered better positioned for the coming energy transition and underweight those less well prepared. They are otherwise relatively unconstrained. Obviously, LCTU and LCTD hope to outperform their broader indices over time as the growing risks of climate change play out. According to Bloomberg data, at least through the end of 2021, they had.

One fundamental question investors must ask themselves is whether these types of ETFs will outperform their non-transition-ready counterparts over longer periods of time. Beyond risks from the "E" in ESG, they must also ask whether other ETFs and new indices similarly designed to limit exposures to social and governance risks will outperform their unscreened counterparts over specific time horizons. This means they must successfully navigate how certain values impact financial valuations over time.

Are you ready for some excitement? I hope so—because the answers that are already emerging are not at all what you might expect.

PART 2
The Perils . . .

Chapter Nine

VALUES VERSUS VALUATIONS

The aim of the investments in the Government Pension
Fund Global (GPFG) is the highest possible return with
acceptable risk. Within this overall financial objective, the
fund is to be a responsible investor. A good long-term
return is dependent on economically, environmentally
and socially sustainable development.

"CLIMATE RISK IN THE GOVERNMENT PENSION FUND GLOBAL,"
LETTER FROM THE NORWEGIAN CENTRAL BANK TO THE
NORWEGIAN MINISTRY OF FINANCE

A fundamental challenge for those who wish to have their values and core beliefs represented in their investment philosophies is the impact those decisions may have on their investment returns. Ethical investing facilitates moral consistency, but it is seldom free; it entails risk. That risk usually manifests itself as "tracking error." *Tracking error* refers to possible divergence from one's stated benchmark or chosen index. This is why we spent all the time we just did on how indices work. One lesser-known fact is those who chose to adopt socially responsible investing principles in recent decades— specifically screening out alcohol, tobacco, gambling, and firearms, a.k.a. "sin stocks"—incurred *negative* tracking error. This means those who embraced sin stock exclusions lost money relative to the broader indices, which included them. Stated differently, sin stocks in recent decades have outperformed the broader market. A recent review of "Christian-based" screened mutual funds in the United States found they, too, as a group, consistently underperformed the S&P 500.[1] To be clear, this does not mean investing in ethical or religious-screened funds is "wrong," not in any normative sense. Investors have the right and perhaps even the responsibility to align their values and belief systems with their investment strategies. It only meant, and means, that screening out certain companies can—and, in these cases did—come at a financial price. This widespread underperformance further suggests that specific fiduciary standards—that is, those that require operating in one's clients' best financial interests—may have been strained if socially responsible investment and Christian-screened strategies were recommended to their purchasers without proper disclaimers about their potential—and in these cases, likely—underperformance.

These results are highly counterintuitive. Getting rid of socially "irresponsible" or "bad" companies was supposed to improve returns, not penalize them. The converse was true. The possible implications of this for many ESG strategies are also clear. A key question we must now ask is, Will the profusion of negatively screened ESG strategies suffer the same costly tracking error as sin stocks? If "bad" energy and utility companies outperform the broader market in the years to come, that could end up being the case. And if other "bad" companies habitually outperform "good" ones, are appropriate disclaimers being made about the potential underperformance of ESG strategies and indices? I am not claiming they will; no one knows. Still, you can see why the question must be raised.

Data on whether ESG strategies to date have generated positive or negative tracking error is mixed. A widely cited academic study of ESG equity fund performance covering 2004 to 2018 found that a majority failed to produce any statistically significant positive or negative gross alpha.[2] A separate, more recent meta-study of more than one thousand research papers on ESG and financial performance conducted jointly by Rockefeller Asset Management and the NYU Stern Center for Sustainable Business found some positive correlation at the corporate level but cautioned against drawing any definitive conclusions owing to inconsistent terminology, data shortcomings, and irregularities across the diffuse range of ESG investment strategies.[3] On the other hand, the American financial service firm Morningstar has reported more than 350 basis points of annual outperformance from 2016–2021 for its "U.S. Sustainable Moat" strategy versus their broad market index—very significant, positive tracking error. They also report that 61 percent of their ESG-screened indices beat their broad market equivalents in 2021, down slightly from 72 percent in 2020.[4] This data is encouraging for ESG because oil and gas stocks vastly outperformed the broader market in 2021; for many other ESG strategies in 2021, the strength of oil and gas stocks meant a meaningful component of their COVID-19 pandemic–induced 2020 outperformance ended up being reversed.[5]

Whether ESG index strategies will under- or outperform broader indices in the future will be known only in the fullness of time. We do not know for sure, nor should anyone claim they do. Most likely, some will and some won't. Definitions are a large part of the challenge. The terminology conundrum found in the Rockefeller–Stern study remains acute. There is no consistent or emerging standard for what constitutes a "sustainable" strategy. Data sets on the widening array of ESG strategies remain too inconsistent to be analytically comparable. Thousands of current investment strategies now labeled "ESG" or "sustainable" lack fundamental conformity. It appears some asset managers are relabeling existing funds as "sustainable" without even changing their underlying investment strategy, a practice that may involve greenwashing.[6] Other studies have shown that strong "G" fundamentals generate outperformance, especially in emerging markets.[7] This shouldn't be surprising, however, nor is it a fundamentally new insight. Better-run companies have historically performed better. They always have, and always will. New ESG metrics aren't needed to tell us that "G" is important. "G" has always been important. As discussed earlier in this chapter and in chapter 7, this is another example of why it may be helpful if "E," "S," and "G" risks were more clearly segmented. When conjoined, each risk factor can muddle the others. "E," "S," and "G" risks should each be given due consideration in systematic risk mitigation models. Each of its own accord may well prove material and simply overwhelm the others.

As the former SEC chairman Jay Clayton rightly observed, statistically rigorous and precise comparisons of companies and industries on *composite* ESG metrics risk being overly inclusive. To say Pepsi is a 17.7 and Coca-Cola a 25.3 conveys no actionable insight into their relative value at any given moment, especially since other ESG score providers may rate them differently. Over time, reported corporate data will become more consistent, ideally along TCFD and SASB standards. As that happens, more rigorous studies and more conclusive evidence will also emerge. Based upon the data that is widely available now, however, one cannot definitively claim that negative screening of specific companies and industries with poor composite ESG metrics will consistently generate positive *or* negative alpha in the years to come.[8] Outcome possibilities skew both ways: toward outperformance and underperformance. Until more rigorous data is available, all institutional investors—including public pension plans, insurance companies, endowments, and foundations—as well as individual investors will need to develop supplemental reasons and methodologies for adopting or avoiding ESG considerations in their investment portfolios. In effect, they are left to estimate how environmental, social, and governance screens may or may not impact their ethical, emotional, and financial well-being. Every investor also needs to decide whether operating with their own normative values and assumptions compensates for the uncertain impact those decisions may have on the companies they choose to underweight or overweight, as well as the valuations of their portfolios over longer periods.

Unsurprisingly, this uncertainty combined with widely divergent ethical priorities has led many sophisticated institutions to different conclusions. For example, the Norwegian central bank, on behalf of Norway's Government Pension Fund Global, recently advised their fund's trustee, the Norwegian Finance Ministry, not to divest from entire industries or to adopt ESG-screened indices. Importantly, they arrived at their recommendation to maintain their long-held, much more inclusive benchmarks because of uncertainties around climate: "We do not believe there is sufficient evidence to claim that climate risk is systematically mispriced."[9] Indeed, rather than outright divestment or attempting to institute portfolio-wide temperature alignment goals, Norges Bank asserts that "climate change is an area that may be well suited to active management." In other words, they believe investment strategies that actively seek out energy transition winners and losers will likely be more rewarding than those that employ negative ESG screening.

Others—including a growing number of university endowments—have come to the opposite conclusion. These sophisticated institutions have pledged to exit all of their fossil fuel holdings and committed to owning portfolios with net zero greenhouse gas emissions before 2050—and in one case, that of the University of Cambridge, by as early as 2038. Harvard, Cambridge, the University of Pennsylvania, the University of Michigan, and others who have taken similar pledges conducted detailed analyses on the potential costs and benefits of screening out certain industries and corporations based upon their specific ESG concerns. While admitting they did not know with certainty how much investing in accordance with the goals and hopes of the Paris Agreement may or may not cost, they still felt compelled to act. As Cambridge's chief investment officer, Tilly Franklin, put it, "Climate change, ecological destruction, and biodiversity loss present an urgent existential threat, with severe risks to humankind and all other life on Earth. The Investment Office has responded to these threats by pursuing a strategy *that aims to support and encourage the global transition to a carbon-neutral economy*" (emphasis added). Cambridge took this decision while acknowledging it was doubtful their divestment would have more than a symbolic impact on the underlying companies. They were certainly correct to surmise this; their endowment was recently valued at $4.5 billion, or 0.0033 the size of the $1.4 trillion Norwegian pension fund that has so far chosen not to divest.

Given that the volume of assets that will not follow Cambridge's divestment strategy is so large, the direct financial impact on any company they may choose to include or exclude is statistically meaningless. They may "aim to support and encourage" the transition, but they cannot do so in any meaningful financial or practical sense.

The Norwegian government pension fund is remarkable in many respects. It was funded from the excess proceeds of Norway's petroleum exports over the past thirty years and is still referred to as the "oil fund" by residents. With a recent market value of around $250,000 for each Norwegian citizen, it is also the largest sovereign wealth fund in the world. Its stated mission is "to safeguard and build wealth for future generations." It seeks to be a stable funding source for future Norwegians long after the country's oil resources have been depleted, ideally in perpetuity. At their peak, Norwegian oil fields produced 3.4 million barrels of crude per day. And even though the government pledges investment in and national commitments to sustainability, Norway is still exporting its oil. Norway's remaining reserves are estimated at eight billion barrels, versus Saudi Arabia's nearly 270 billion. Norway seems rather eager to dispense all its oil relatively quickly. In addition to providing for its own needs, Norway supplies about 2 percent of oil demand globally. Norway is also the third-largest exporter of natural gas in the world, behind only Russia and Qatar. While there is an ongoing review concerning their investment principles and practices, it is also fair to say Norwegians do not suffer from excessive introspection over their country's direct and verifiable contributions to global greenhouse gas emissions. "In Norwegian politics, there's been a very successful attempt to separate the discussion of oil policy from the discussion of climate policy," Bård Lahn, a researcher at Norway's Center for International Climate Research in Oslo, reports.

Unlike most other institutional investors, the Norwegian oil fund is also remarkably transparent about its holdings. Details of nearly every asset it owns can be found on its website. Given its size and broad strategy, it now owns stakes in more than nine thousand companies in seventy countries worldwide, maintaining an average position of 1.5 percent in each name. You'll recall that John Bogle advocated owning the "whole haystack" rather than trying to find the needle within. Bogle correctly deduced that the broader one's investment universe, the more likely it is to track global growth. The Norwegian oil fund mindfully practices Bogle's haystack principle on a global scale.

Norway's oil fund recently outlined its commitment to investing sustainably and exercising responsible share ownership in a detailed white paper on responsible investment.[10] As long-term investors, they know economic growth must be sustainable, that markets must function efficiently, and that the many companies they invest in must practice good governance. Norway's oil fund also publishes expectations of the companies it owns on various topics, including board independence, unequal voting rights, share issuance, and sustainability reporting. On behalf of the Norwegian people, the oil fund practices an engaged stewardship model, voting in thousands of shareholder meetings every year—including more than 120,000 shareholder resolutions in 2020 alone. Diligent with data, they further monitor the carbon footprint of their holdings. That footprint isn't negligible. In 2020, the fund estimated that its equity holdings accounted for 92.4 million tons of CO_2 equivalents. It does not report whether this is too high, too low, or just right, nor does it publish any target for what it wants this amount to be in the future.

In explaining its recommendation not to align the fund's portfolio holdings with the Paris Agreement or to impose more climate-related restrictions on its investment universe, the Norwegian central bank has noted its potential inconsistency with its assigned mandate. That mandate requires it seek the highest possible return with acceptable risk—in other words, a typical fiduciary standard. Officers for the Norwegian oil fund also argue very persuasively that the best way to change corporate behavior is to remain an active shareholder:

> A climate-adjusted index will not include companies or sectors that currently fail to meet given climate-related criteria but may play a role in the climate transition in the future. A decision to replace the current equity index with a climate-adjusted index would mean the fund misses out on opportunities where companies not included in the index undergo a transformation. Not investing in companies or sectors by using a climate-adjusted index is not a very suitable tool for bringing about changes in corporate behavior.
>
> Through our contributions to standard-setting, clear expectations, dialogue with companies and voting, we will seek to ensure that the companies in our portfolio are well equipped for the low-carbon transition.[11]

If the Norwegian central bank's advice is followed, it appears that the largest sovereign wealth fund in the world—Norway's oil fund—will retain maximum latitude to invest as it pleases to maximize returns across all industries, including carbon-based energy and mineral extraction companies, consistent with their historic risk guidelines. The central bank's report explicitly rejected the idea of investing in any manner relating to Paris targets or constraints. For this reason, it is instructive to juxtapose the bank's conclusions with other sophisticated institutions that have taken opposing positions, by either explicitly committing to Paris-aligned goals in their investment guidelines and strategies or adopting other temperature alignment constraints. Consider, for example, the positions of these five university endowments, all deeply respected. Note in particular, the "virtue competition" that seems to be taking place between Cambridge, Massachusetts (where Harvard is located), and Cambridge, UK.

HARVARD UNIVERSITY

The President and Fellows of Harvard College instructed the Harvard Management Company (HMC) to set the Harvard endowment on a path to achieve net-zero greenhouse gas (GHG) emissions by 2050. This pledge is a first among higher education endowments and a natural extension of Harvard's ongoing efforts to prepare for and accelerate the necessary transition to a fossil-fuel-free economy. (April 2020)

UNIVERSITY OF CAMBRIDGE

The University of Cambridge aims to divest from all direct and indirect investments in fossil fuels by 2030 as part of the University's plan to cut its greenhouse gas emissions to zero by 2038. This plan puts Cambridge at the head of the race to become the first university endowment of its kind where greenhouse gas emissions from the activities of all investments balance out at zero. (October 2020)

COLUMBIA UNIVERSITY

Columbia University will combine its formidable strengths in scientific research and policy with the skills of its investment team to play a constructive role in a broader shift in the global economy to net zero emissions by 2050. Columbia University will not make any new investments in private funds that primarily invest in oil and gas companies. (January 2021)

UNIVERSITY OF MICHIGAN

The University will immediately shift its natural resources investments to focus more on renewable energy, stop investing in funds primarily focused on certain fossil fuels and discontinue direct investments in publicly traded companies that are the largest contributors to greenhouse gases. These new investment strategies are part of a commitment to ensure the University's investment portfolio reaches a "net-zero" carbon footprint by 2050. (March 2021)

UNIVERSITY OF PENNSYLVANIA

Our goal is to eliminate the anthropogenic greenhouse gas emissions footprint created by our endowment's underlying investments (by 2050). Penn's goal is consistent with the 2015 Paris Agreement and the UN's Intergovernmental Panel on Climate Change. (April 2021)

It's worth noting that none of these announcements was accompanied by compelling evidence or new studies suggesting higher returns were likely to result from these new strictures. In fact, in the most detailed report accompanying these carbon commitments—that of Dr. Ellen Quigley, Emily Bugden, and Anthony Odgers for Cambridge—the authors first state, "There is little evidence to suggest that a global portfolio invested to exclude fossil fuels would underperform one that included them." However, they later note there could well be other adverse impacts: "A policy of full divestment would nevertheless necessitate a change in the University's investment model"—a model that relies heavily upon external active managers. Foregoing this form of active external management could be expensive for Cambridge, according to Quigley, Bugden, and Odgers: "Applying the historic 1.2 percent annual outperformance to the [endowment's] entire value . . . would imply a reduction of £40 million per year of investment returns."[12] For Cambridge and most other institutions of higher education, £40 million per year is a significant sum of money. This amount per year could fund large amounts of research into new carbon-reduction technologies or scholarships for underprivileged students. But such potential lost income still did not dissuade their overseer's decision to move to a portfolio with goals of net zero greenhouse gas emission—by 2038, twelve years earlier than the 2050 Paris objective and their Cambridge, Massachusetts, counterpart. At Cambridge, UK, normative values trumped valuation concerns.

While the University of Cambridge and its fellow institutions made commitments to constrain their investment universes, it is also important to note that none who committed to a net-zero investment framework unveiled the tools or methodologies it plans to use to measure its portfolio's carbon

footprint verifiably. Aside from incomplete corporate data, vexing accounting challenges like netting shorts versus longs and the measurable contributions of existing assets like arable farmland must still be settled. Why are
these methodologies so important? If all one needed to achieve a net-zero
portfolio was to sell short oil and gas stocks or other large carbon emitters,
one could claim compliance relatively simply. Similarly, buying existing forests or farms might be another way to improve one's net carbon footprint.
Unfortunately, however, neither strategy would actually result in lower net
greenhouse gas emissions; both are merely carbon accounting measures.
What should be targeted instead is *additionality*—that is to say, improvements that wouldn't have occurred but for specific investments. Many other
outstanding carbon accounting practices need to be debated before best
practices are ultimately agreed upon, too. Committing to a net-zero portfolio
target today is somewhat like agreeing to a gluten-free diet before knowing
which foods contain gluten or what nutrients one might be sacrificing in the
process. It may confer a certain peace of mind and be welcomed by a number
of vocal critics, but it should not be lauded for any practical contribution to
enhanced environmental sustainability. Not yet anyway.

Perhaps for these reasons, the retention of significant investment latitude
remains a clear priority for other university endowments. One of those is
the Stanford Management Company (SMC). In a recent restatement of their
"ethical investment framework," Stanford's endowment team clarified how
and why they would not follow Cambridge's approach:

> As a prudent fiduciary, SMC incorporates the risks associated with carbon
> when considering conventional energy holdings. In economic terms, we try
> to account for the externalities associated with burning hydrocarbons, which
> helps us invest sensibly in a sector undergoing significant change. While cer-
> tain ethical and social risks rise to the broad level of asset allocation, many
> risks are best analyzed at the level of specific businesses. The businesses in
> our portfolio provide highly valuable goods and services to the world. We
> believe that well-run companies which respond to genuine consumer needs
> in a responsible fashion have a beneficial impact on society.
>
> We consider a strong moral framework to be an alignment of interests with
> the University. Nevertheless, while we ask our investment partners to dem-
> onstrate strong moral sensibility, it would not be appropriate to insist they
> advance a particular social or political agenda for its own sake. Such insistence
> would contravene a guiding tenet of Stanford as an educational institution: to

avoid taking political and ideological stances. SMC is obliged to place proper weight on ethical issues that can have a bearing on economic results but not to use the endowment to pursue other agendas.

Like Stanford, Yale University has also directed its asset managers to respond to the growing climate crisis conundrum thoughtfully, while stopping short of divestment. A committee advising their endowment recommended that their exposures to firms that engage in fossil fuel production be subjected to five principles:

1) Fossil fuel producers should neither explore for, produce, or supply fossil fuels, nor engage in extraction methods that result in high GHG emissions relative to the energy supplied, if there are feasible alternatives that result in significantly lower GHG emissions.
2) Fossil fuel producers should operate in a manner consistent with best industry practices to reduce GHG emissions.
3) Fossil fuel producers should not undermine but support sensible government regulation and industry self-regulation addressing climate change.
4) Fossil fuel producers should not undermine but support accurate climate science and public communication about fossil fuel products, climate science, and climate change.
5) Fossil fuel producers should be transparent with Yale and Yale's Investment Managers about their compliance with Principle Nos. 1 through 4.

While there is no broad energy divestment or Paris-aligned deadline at either Stanford's or Yale's endowment office at this writing, strong forces are still agitating for them. Active campus debates on the relevance of aligning one's investments with a net-zero glide path will doubtless continue. What activists in these debates should ask is, What verifiable impact would Paris-alignment by their university's endowment actually have?

The decision by Stanford, the Norwegian oil fund, and others to retain maximum investment flexibility—refusing to disallow oil, gas, or any other industries, companies, or strategies that may not conform to the demands of the Paris Agreement—enables their active managers to invest their assets by taking strategic, long-term stakes in many more companies. It also allows them and their external partners to take *tactical*, short-term positions in names and sectors that may trade attractively from time to time—like when they are inexpensive relative to the market overall or within a sector. For example, when oil and gas prices moved higher at the beginning of 2021 owing to excess oil demand versus supply, many oil stocks like Chevron's rallied sharply versus the market overall, creating a significant alpha-generating opportunity relative to the broader market (figure 9.1). In fact, the energy sector registered more than a 60 percent return during 2021 versus a 27 percent gain for the S&P 500 overall, and it outperformed even more in the first quarter of 2022. Similarly, after significant shareholder resolutions championed by Engine No. 1 required Exxon to replace three of its board members and to be more attentive to investment opportunities in the renewable energy space, Exxon's stock outperformed its competitor Chevron, as shown in figure 9.2. Blanket industry exclusions, such as those embraced by Cambridge, will preclude relative value investment opportunities like these from benefiting endowments and their missions. Investment arbitrage opportunities like these are precisely why Norges Bank concluded that the global

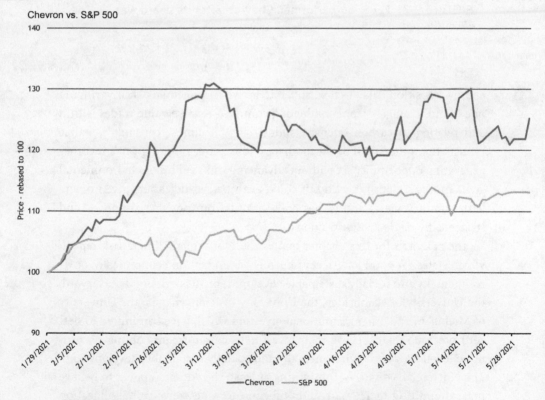

FIGURE 9.1. Chevron outperformed the S&P 500 early in 2021.
Figure by the author.

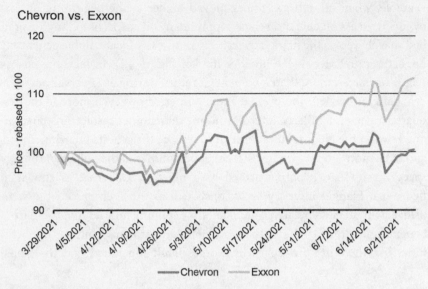

Chevron vs. Exxon

FIGURE 9.2. Exxon outperformed Chevron after its board was reconfigured.
Figure by the author.

energy transition "may well be suited to active management." Hamstringing one's fund managers from conducting common relative value trades within and relating to carbon-intensive industries will almost certainly penalize returns that others will be all too happy to exploit. Limiting one's income-generating opportunity set without advancing one's values or environmental goals merits deeper reflection. It could even be self-defeating, given others with contrary views and values could profit from one's self-imposed and strategically ineffective restrictions.

The epigraph for this chapter quotes the Norwegian climate risk report, which states, "A good long-term return is dependent on economically, environmentally and socially sustainable development." It is evident that Harvard, the University of Cambridge, the University of Pennsylvania, the University of Michigan, and other institutional investors which have committed to net-zero portfolio goals believe their investment commitments ultimately promote "economically, environmentally and socially sustainable development." The Norwegians differ. At this writing at least, they do not appear to believe that committing their $1.4 trillion oil fund to a net-zero capital allocation decision over any period is particularly critical to promoting economically, environmentally, and socially sustainable development. To the Norwegians, other efforts would better promote these goals. They summarize their sustainable investment commitments in a 2020 responsible investment white paper, relating their argument to "acceptable" investment risk:

> We see opportunities in companies that enable more environmentally friendly economic activity. We identify long-term investment opportunities by analyzing companies' operations and the impact they have on the climate and the environment. We divest from companies where we no longer wish to be a shareholder for ethical or sustainability reasons. By not investing in these companies, we reduce our exposure to unacceptable risks.

As mentioned, the Norwegian oil fund does have some ethical exclusions. The fund does not invest in companies that produce tobacco, manufacture weapons that "violate fundamental humanitarian principles," or use or produce a specific amount of thermal coal. In a well-documented and high-profile move in 2006, moreover, the Norwegian oil fund also famously divested from Walmart, the largest private employer and retail chain in the United States. They did so owing to "serious, systemic violations of human rights and labor rights." Given Norway is a major oil producer and exporter, however, the fund somehow sees the energy industry differently.[13]

As we have seen, the dynamic trade-off between individual and full enterprise beliefs and asset valuations creates vexing considerations and concerns that do not lend themselves to formulaic resolution. Cambridge has elected to proceed with a commitment to align its portfolio to the Paris Agreement on a fast track even though it recognizes that it may cost the endowment £40 million in foregone revenues per year. If the Norwegians applied the Cambridge policy to their oil fund, and if it had the same relative impact, it would cost about $17 billion per year. For context, $17 billion is about ten times the amount the European Union has allocated for humanitarian aid in 2022.

But values and valuations need not always be in conflict. When we turn to our detailed discussion of impact investing, in fact, we will discover multiple examples in which values and valuations are entirely consonant. In these cases, however, the underlying investor must make an explicit decision to pursue outcomes that may deviate significantly from commonly used benchmarks and to prioritize objectives beyond risk-adjusted returns. Impact investors explicitly commit to "doing well and doing good." Unless and until "double-bottom-line" investment objectives like those commonly found in impact investment strategies are embraced, values and valuations will often involve trade-offs. Over extended periods, these trade-offs may prove to be very financially material. When one knows that an ethical or values-based decision may create negative tracking error, every effort must be made to ensure potentially forgone income is compensated for by attaining one's other desired goals, whatever those may be.

Investment risk, whenever it must be taken, should be intentional, diversified, and properly sized. Risks that aren't deliberate are often the most painful over time. Unfortunately, they are also often the hardest to detect. This is why SASB, TCFD, and so many others are working intensively to surface undetected material risks and why they, along with many global regulators, activists, and official certification efforts are endeavoring to do whatever they possibly can to help systematically "hardwire corporate goodness." Achieving optimal levels of inclusive, sustainable growth through heightened corporate mindfulness is a goal worth striving for. It's also what we collectively committed ourselves to at this book's outset, as well as its central purpose. What roles should business and finance play in helping to fashion the good society? And if positive ESG influences are not yet as pervasive as we had hoped, how else might investors effectuate more inclusive, sustainable growth with their capital allocations, actions, and beliefs?

It turns out there are many strategies, programs, and techniques other than those in the ESG movement which hope to promote corporate "goodness." We should spend some time exploring them and highlight those with the greatest prospects for success.

Chapter Ten

HARDWIRING CORPORATE GOODNESS

Goodness is the only investment that never fails.
HENRY DAVID THOREAU

Let's recap.

For ESG investing to work as many hope and expect—that is, as investment vehicles that systematically promote more inclusive, sustainable growth while simultaneously generating superior returns—continued progress is needed in a multistaged yet broadly interdependent sequence.

To begin with, recognized and universally followed standard setters like SASB and TCFD must continue to identify and set the right objectives for which corporations must strive to achieve. If they prioritize the wrong material risks or set the wrong methodologies for measuring them, their essential role in directing corporations to make further progress on a range of socially desirable objectives, including the UN's Sustainable Development Goals, will be compromised.

As these standardized reporting benchmarks are being updated and improved, corporations must simultaneously become more diligent about adopting them, working deliberately toward reducing specified material risks, and employing best environmental, social, and governance practices. They must follow these new standards consistently, comparably, and transparently, moreover. If they do not strive to conform to standards like those of SASB and TCFD, or if they try but do not report their efforts accurately, many indexed ESG investment strategies and the research upon which they depend will continue to generate random returns. No strategy can be any better than the data upon which it is based. Given Morningstar reports that

U.S. corporations disclose only two-thirds of their genuine, material ESG risks, versus 73 percent in Europe and 52 percent in Asia, when it comes to complete corporate reporting, there is still a long way to go.

Finally, capital markets and investors must reward valid ESG progress proportionately. For this to happen, bankers and investors must identify and favor the best ESG corporate actors accurately relative to their specific behaviors. They need to do so across every industry that is part of the economic growth value chain, moreover—not just cleantech and circular exemplars, but also steel manufacturers, cement makers, agricultural giants, and energy providers. Why? Economic growth will continue to depend on many carbon-intensive industries. There is no viable path forward for optimally inclusive growth through a handful of zero-carbon-emission industries alone. *Net* zero depends upon netting, after all. In other words, once every viable minimum-carbon process has been adopted, we will still need to find economical methods for removing unavoidable carbon emissions from the atmosphere. After all, we will still be using some carbon-dependent processes and fuels. By providing patient and loyal capital discernably, investors will ultimately assign higher equity valuations and lower debt costs to the paragons of each industry. This relative valuation process is needed to complete the virtuous feedback loop upon which the systematic promise of most ESG investing depends. If ESG paragons don't outperform laggards, indexed ESG strategies will not outperform their unscreened counterparts. In short, for ESG investing to work, verifiable and verified corporate "goodness" must be financially rewarded consistently. For progress to be systemwide, moreover, "bad" corporations need to believe it is in their interest to adopt TCFD and SASB standards and seek improved ESG scores because they believe it will lead to their financial affirmation as well as socio-environmental progress. If they do not consider it to be in their interests or in the interests of their long-term shareholders, they are highly unlikely to do it.

Clearly, the road to hardwiring corporate goodness I have just spelled out will be a long and winding one. It entails multiple interdependencies. ESG investment outperformance is possible, perhaps even likely, as data improves. Its ultimate success is also in our best interest. We should all want the broader ethos of ESG to reign. That said, its success is certainly not guaranteed, and its reliability is not imminent. The interdependent sequence just described could easily break down.

At current pace, we are years away from the broad adoption of TCFD and SASB standards. Over this period, multiple ESG standards will continue to evolve, especially "S" mores. As a result, properly discounting all knowable ESG risks will be time-consuming, complex, and fraught. Until greater transparency and comparability are achieved, investors must rely upon at least two other divining rods: (1) the types of well-intended, improving, yet still imperfect metrics created by professional rating agencies like Sustainalytics, S&P, MSCI, and others; and (2) other proprietary research and techniques that will differ dramatically from adviser to adviser. As we've seen, no comprehensive ESG rating methodology can be any better than its raw data inputs. Until more consistent, comprehensive corporate data becomes available, ESG rating agents will depend upon large amounts of subjectivity. Investors must make crucial relative value decisions based upon these and their own analytically imprecise methods. This is not to impugn any of these efforts; it is only to make clear their limitations.

Recognizing the challenges ESG rating agencies face and the opportunity to innovate, it is no surprise several alternative professional efforts are now underway to identify and reward "good" companies systematically. Most of these initiatives claim to have found unique, reliable methodologies for identifying companies that generate large amounts of shareholder value through greater stakeholder attentiveness. Happily, some of these alternative methodologies have shown considerable promise. In the context of our quest for more inclusive, sustainable growth, it is worthwhile to survey these alternative, non-ESG initiatives. One of the more creative was started by the legendary investor Paul Tudor Jones.

It's been said that saints are sinners who just kept going. This seems a particularly apt aphorism for Paul Tudor Jones. For much of his career, the billionaire Tudor Jones was the type of successful hedge fund executive who environmental and social justice activists might love to hate. Part of their disdain might stem from his personal actions. In 1990, Tudor Jones pleaded guilty to illegally destroying eighty-six acres of protected wetlands on his Maryland Eastern Shore hunting estate. Another source of concern might be professional. His highly successful global macro fund, Tudor Investment Corporation, thrived on short-term bets involving rates, derivatives, and commodity contracts—that is, things that have little to do with providing patient capital to the real economy. Moreover, in 2012, he bought the energy trading business of Louis Dreyfus and Highbridge Capital, renaming it Castleton Commodities International. Instead of just flipping financial contracts, Castleton made Tudor Jones a significant player in the physical delivery of petroleum and natural gas, with considerable financial interests in carbon-based fuels. No longer just rich, he became filthy rich, as it were.

But somewhere along this seemingly wayward journey, Tudor Jones appears to have experienced his own genuine "Road to Damascus" moment. In 1988, along with Peter Borish and Glenn Dubin, he founded an exceptional organization: the Robin Hood Foundation. The Robin Hood Foundation today is widely recognized as one of the most impactful philanthropic endeavors in history. It has raised and donated hundreds of millions of dollars to alleviate poverty and its causes, most especially in and around the New York metropolitan area. As its name suggests, it excels at taking money from the rich and giving it to the poor. Much of that money came from Tudor Jones himself.

But Paul Tudor Jones's commitment to using his significant personal resources to aid those who have been left behind certainly did not end with Robin Hood. In 2013, he led a group of highly motivated individuals—including Deepak Chopra, Arianna Huffington, and Paul Scialla, a notably diverse crew—to establish a new nonprofit organization, Just Capital. The mission of Just Capital is "to build a more just marketplace that better reflects the true priorities of the American people." It is based upon the genuine belief that capitalism, for all its flaws, can still be a positive force for change if more information about what U.S. corporations are doing with, for, and to all their stakeholders could be brought to light. As a nonprofit outfit, they have no direct financial interests in or from their work. Instead, the primary benefit they hope to effect is a more inclusive, sustainable economy that works for all, as well as greater corporate alignment with our evolving moral priorities.

The primary activities of Just Capital are proprietary research and data collection exercises that aim to measure corporate impacts on their multiple stakeholders relative to the evolving social priorities of Americans. Specifically, Just Capital systematically quantifies seven major focal areas among the top one thousand U.S. companies: (1) treatment of workers; (2) treatment of customers; (3) nature of products; (4) impact on the environment; (5) impact on communities; (6) impact on the labor force; and (7) the nature of leadership. Just Capital also conducts an extensive annual survey of Americans, asking them to rank their most significant concerns—things like the environment, diversity, wage fairness, and corporate profitability. Companies are then rated by these evolving priorities overall and within their industrial subcategories. Thus, as American priorities shift, Just Capital's corporate rankings shift. For example, in their 2020 survey,

Chevron was ranked the "most just" company in the oil and gas segment and the sixty-first most just company overall, MetLife was the second most just among all insurance companies and seventy-fifth overall, and Goldman Sachs was fifth within the capital markets segment and one-hundred-thirty-second overall. The year before, ConocoPhillips outranked Chevron, MetLife was forty-forth overall, and Goldman Sachs was a distant one-hundred-sixty-fourth. Note, however, that these changes may have had as much or more to do with evolving American concerns than with changes within the corporations mentioned. The considerations of whether they pay a living wage, support work–life balance, use resources efficiently, pay taxes, provide safe working conditions, or overcharge for their products and services were all carefully measured. They were then weighted relative to what polled Americans said they most value.

"Over the past fifty years," Tudor Jones says in a TED Talk that has now been watched more than two million times, "we have come to view our companies and corporations in an almost monomaniacal fashion. We have emphasized profits, short-term earnings, and share prices to the exclusion of all else. It's like we ripped the humanity out of them."

But shining a light on the amount of humanity there is within individual American businesses is not all Just Capital is trying to do. They also believe corporations that genuinely strive to align their activities with the American public's social priorities will also financially outperform their less enlightened counterparts over time. To this end, Just Capital combined forces with Goldman Sachs Asset Management to launch an actively managed ETF, wittily listed under the ticker symbol "JUST." Much as Just Capital ranks the top one thousand U.S. companies by their proprietary metrics, the asset management arm of Goldman Sachs strives to reweight Just Capital's companies similarly in the ETF, following Just Capital's annual surveys.

Forbes magazine reports the results of Just Capital's proprietary work annually in a compelling article often titled "America's Most Just Companies." It is the kind of award every company would covet highly if they won— and ignore if they did not. The top five most just companies in 2020 were all household names: in order, Microsoft, Nvidia, Apple, Intel, and Alphabet. Five others—Texas Instruments, IBM, AT&T, Merck, and Salesforce— had also scored at or near the top in previous years. Because of their low carbon footprints, tech firms, health care companies, finance, and specific retail, food, and beverage firms usually score more highly. General Motors was the most highly ranked industrial corporation in the 2020 survey, at 28.

Dominion Energy was the highest-ranked utility, at 64. As just or unjust as these companies may seem to be, it's impossible to imagine how Americans could live as they now do without the forms of transportation and electricity companies like GM and Dominion provide.

Just Capital has been notably transparent about the limits of their approach. In a detailed case study published by the Harvard Business School, Just Capital's director of indices and analytics, Hernando Cortina, states, "Just Capital is built upon the premise that the data are always going to be evolving. Measurement is imperfect. We will get better and incorporate new data sets as they become available. But we knew from the start that . . . it's going to take years to perfect this process."[1] In many ways, Just Capital's challenge is no different from that faced by SASB, TCFD, and ESG investors more broadly: they, too, need corporations to report all the data needed to make better normative and financial judgments.

According to Bloomberg data, Apple, Microsoft, Amazon, Alphabet, and Nvidia remained the Just ETF's top holdings during much of 2021, whereas Devon Energy and Under Armour were much smaller allocations. Interestingly, the market capitalization of the Just ETF has failed to keep pace with the broader market since launch, suggesting initial investors may have lost confidence in the strategy. This could be because of its relatively high fees or a growing conviction that Goldman Sachs Asset Management's and Just Capital's ranking methodologies are not producing attractive risk-adjusted returns. Bloomberg reports that from the time of its launch through the end of 2021, the Just ETF underperformed the S&P 500.

Just Capital employees are capitalists committed to maximizing the good public U.S. businesses do by transparently ranking how they treat all their stakeholders versus their peers and one another. This would be an excellent service to socially responsible investors if they could. The combined economic impact of the U.S. private sector is more than four times the size of government programs in the United States and more than forty times the size of annual philanthropic gifts. Just Capital rightly claims it will not be possible to create a country of greater shared prosperity and sustainability without aligning U.S. businesses more toward stakeholder needs. Note, however, that no company went to Paul Tudor Jones and his cofounders asking them to create a new methodology to assess and rank their innate goodness. Just Capital started their activities without solicitation or industry backing, intending it simply as a method for bringing more "humanity" into the corporate world. And while many corporations actively engage Just

Capital researchers for more favorable consideration, they can't and don't pay Just Capital to be a part of their surveys or to be ranked more highly. Corporations that wish to seek special accreditation for their model behavior can consider another nonprofit organization that measures and monitors social contributions rigorously, however. One of the most important of these accreditation programs is B Lab. If a corporation's accreditation application to B Lab is successful, it can achieve an increasingly coveted title: Certified B Corporation.

According to the B Lab website, Certified B Corporations meet the highest standards of verified social and environmental performance, public transparency, and legal accountability through deliberate managerial efforts that most effectively balance profit and purpose. Many in the vibrant conscious capitalism movement agree with this assessment. By the end of 2021, more than 3,500 corporations in seventy countries and 150 industries had achieved B Corp certification. Another one hundred thousand globally are said to have referenced the detailed methodologies and impact assessments B Lab has created. That's a lot of impact, and it's global, unlike that of Just Capital.

Along with the United Nations Global Compact, B Lab has also developed the "SDG Action Manager." This professional self-assessment tool has helped tens of thousands of businesses chart actionable paths for achieving one or more of the UN's Sustainable Development Goals through their corporate actions. More than eight hundred B Corps have also pledged to be net zero by 2030, twenty years earlier than called for by the Paris Agreement. They have committed to doing so in part by following B Lab's respected *Climate Justice Playbook for Business V1.0*. This comprehensive playbook was developed in conjunction with several COP 26 Climate Champions, including Provoc and the University of Oxford.

Given these extensive efforts and programs, B Lab may be the most successful *global* effort to hardwire more goodness into the business world. Executive management and boards of B Corporations publicly vow to live by four core values:

- To be the change they seek in the world
- To conduct their activities as if people and place matter
- To do no harm while acting for the benefit of all through their products, practices, and profits
- To behave knowing humans are dependent upon one another and responsible for future generations

At a macro level, the success of the B Lab movement depends crucially upon the desire of more and more corporations wanting its certification. Frankly, a strong case could be made for it. Ultimately, people will likely want to buy from, work for, and invest in the businesses they most trust and admire. If B Corp certification is seen as a meaningful way to build credibility and long-term value for one's business—which it increasingly is—the B Lab movement could meaningfully move the needle toward more inclusive, sustainable growth.

As you would expect of any certification of this stature, the process of gaining B Corp accreditation is demanding. Fewer than one of five who take B Lab's initial assessment even choose to apply. Of those, only about one of three are ultimately approved, meaning a completion rate of less than 7 percent. Because of overwhelming interest moreover, the application and review process can take anywhere from six to twelve months to complete. Winning B Corp certification is time-consuming and intensive, in short. Being a B Corp is not a "one and done" process, either. Certification must be renewed every three years by scoring a minimum number of points versus an ever-evolving checklist and scale.

Whereas all Just Capital corporations are publicly traded, very few Certified B Corps are. The vast majority of Certified B Corps are small unlisted businesses or sole proprietorships. A handful are subsidiaries of publicly traded companies, like Ben & Jerry's and Sundial Brands. Others are well known, like Patagonia and Eileen Fisher. A few Certified B Corps went public after achieving certification, SilverChef and Laureate Education among them. To date, however, few large-scale multinational corporations have attempted to apply for B Corp certification; the process is simply too demanding, and success is highly unlikely. Danone North America—a food-processing company that produces yogurt, baby food, and medical nutrition products—is among the largest Certified B Corps to date. Its annual revenues are a bit below $150 million, though, making it smaller than any corporation in the S&P 500.

Given that applying for B Corp certification entails multiple strategic and logistical challenges, it is unlikely many large companies will try. For these reasons, it is highly improbable that the B Corp movement as currently constituted will attract very many publicly listed companies. From a global welfare perspective, it may make sense for B Lab to tier their accreditation programs to encourage more large corporations to apply. In the absence of such tiering, it seems clear the B Lab model will remain unscalable—something many advocates for greater corporate mindfulness will lament.

At this writing, no other serious, independent, inclusive, and systematic efforts other than those of Just Capital and B Lab seem to be taking root as a means for genuinely hardwiring more ubiquitous goodness into the business world—beyond the ESG investment movement, that is.

I choose the words *serious*, *systematic*, and *inclusive* advisedly. Of course, there are dozens of polls and beauty contests, other aspiring NGOs and not-for-profits, and multiple employment recruitment sites attempting to add to the debate and substance of more purposeful business. On the margin, some of these are also making a difference. For example, Great Places to Work specializes in evaluating and improving employee experiences. It claims to have helped improve corporate cultures in multiple industries across more than sixty countries around the globe. Similarly, and as its name suggests, Glassdoor has given a public window into how employees feel about their employers, a crucial stakeholder concern. But Glassdoor's process is certainly not rigorous in collecting all relevant data; instead, it's an aggregation of opinions offered only by those who bother to write them, meaning it suffers from selection bias.

Elsewhere, a growing group of international corporations has banded together voluntarily and taken a much more demanding oath than B Corps ascribe to. As they are called, these GameChangers promise to cherish the environment, share best practices among themselves, and collaborate where collective efforts would promote a thriving world economy.[2] More than five hundred corporations have taken this pledge "because the profit-at-all-cost model just isn't working." No one comprehensively audits whether their actions align with their words, however. That said, GameChangers appear to have gone several steps beyond the Business Roundtable.

As you'll recall from chapter 4, the Business Roundtable raised many hopes when they issued their detailed 2019 "Statement on the Purpose of a Corporation." Hundreds of CEOs committed to supporting their customers, employees, suppliers, communities, and shareholders equitably while stopping short of any explicit embrace of net-zero goals. BRT corporations collectively employ about fifteen million workers, or about one-tenth of the U.S. labor force. The American Sustainable Business Council is another voluntary society. It represents more than 250,000 small U.S. businesses. They also express their support for and commitment to the BRT'S 2019 statement. Collectively, they employ about eighteen million workers, 20 percent more than the BRT's signatories.

Much like *Forbes*'s partnership with Just Capital, *Fortune* has allied with partners at the Shared Value Initiative to identify successful companies they feel are doing the most to tackle society's unmet needs. The Shared Value Initiative is a significant effort. Since 2015, with their counsel, *Fortune* has assembled and published a "Change the World" list, shining a light on those firms with discernible social impact owing to activities embedded in their core business strategies. Four variables dominate their evaluation and selection process: (1) measurable social impact; (2) business results; (3) degree of innovation; and (4) corporate integration. To garner *Fortune*'s highest "Change the World" rating, the reach, nature, and durability of a given company's impact on one or more societal problems must be clear. In addition, their socially impactful work must also contribute to that company's profitability and shareholder value beyond simple reputation. Finally, all these accruing social benefits must be integral to the company's overall strategy and sufficiently innovative within its industry and among its peers to truly differentiate itself.

The Shared Value Initiative is an outgrowth of the consulting firm FSG, started in 2000 by the Harvard Business School professors Michael E. Porter and Mark Kramer. The Shared Value Initiative took its direction from a famous essay by Porter and Kramer published more than a decade ago, well before ESG investing burst onto the scene.[3] The Shared Value Initiative hopes to build a community of leaders united in making the earth "a more equitable, healthy and sustainable place." Their *Purpose Playbook* helps corporations reimagine their strategies to create "lasting, scalable impact for both society and their business." The Shared Value Initiative believes success can be achieved by leading with purpose, designing a shared value strategy, embedding that strategy into company operations, and empowering talent to reimagine how their responsibilities might be aligned to support it (figure 10.1). The positive feedback loop between purpose and profits is one that the Shared Value Initiative both expects and strives to verify.

Unsurprisingly, *Fortune*'s 2020 "Change the World" list was dominated by firms that led the response to COVID-19, including the vaccine makers Pfizer, AstraZeneca, BioNTech, and Sanofi, all of whom tied for first. Regeneron Pharmaceuticals and Serum Institute of India help round out the top ten. Other names included in the top rankings of 2020 include Alibaba, PayPal, Nvidia, BlackRock, Zoom, Safaricom, and Walmart.

Among those persuaded by FSG's Shared Value Initiative is Harvard Business School professor George Serafeim, a frequent author and the

PORTFOLIO OF PRACTICES FOR A PURPOSE-LED COMPANY

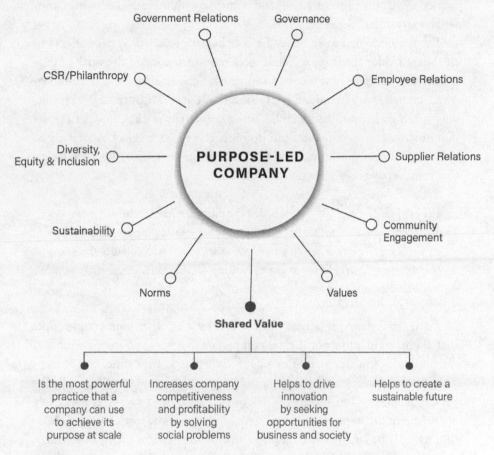

FIGURE 10.1. How the Shared Value Initiative sees purpose and profit.
From *Purpose Playbook: Putting Purpose into Practice with Shared Value,* edited by Georgina Eckert and Bobbi Silten. FSG and Shared Value Initiative, Version 1.1 of the Purpose Playbook was Published in May 2020.

aforementioned authority on ESG matters. "It seems clear companies will be under growing pressure to improve their performance on ESG dimensions in the future," Professor Serafeim recently wrote. "The challenge for many corporate leaders is that they aren't sure how to do it."[4] Serafeim has plenty of ideas of what they should and should not do, meanwhile. "Not all ESG issues are created equal—some matter more, depending on the industry," he says. Which ones? The most material issues, of course. "The only way to outperform in this new era will be for companies to make material ESG issues central to their strategy and operations, to go above and beyond their competitors, and then to measure and communicate their superior performance. . . . If companies are bold and strategic with their ESG activities, they will be rewarded," Serafeim writes.

The methodologies used and the conclusions reached by proponents of the Shared Value Initiative will find further confirmation in the work of Professor Alex Edmans of the London Business School. Edmans combines rigorous analysis with the broad conviction that corporate purpose and profits very often go hand in hand. In his most important work, *Grow the Pie*, he demonstrates how companies can do both good and well not by striving to maximize profits but by striving to maximize their social value. He likens this mindset to a "pie-growing mentality":

> The pie-growing mentality stresses that the pie is not fixed. When all members of an organization work together, bound by a common purpose, and focused on the long term, they create shared value in a way that enlarges the slices of everyone—shareholders, workers, customers, suppliers, the environment, communities, and taxpayers.[5]

Aside from these initiatives, there is at least one other high-profile effort that claims to identify world-class companies that serve all their stakeholders optimally, translating into excellent investments over time. It is impossible to say that this effort is rigorous in any objective sense, however. Its principal author—Raj Sisodia of *Conscious Capitalism* fame—has effectively claimed that his stamp of approval may be the ultimate stock recommendation tool. He has published data (table 10.1) that he claims provides evidence that his so-called firms of endearment have massively outperformed the

TABLE 10.1
One kiss from Raj Sisodia and you're golden—or so he says

Cumulative performance	3 Years	5 Years	10 Years	15 Years
U.S. firms of endearment	83%	151%	410%	1,681%
International firms of endearment	47%	154%	512%	1,180%
Good-to-great companies	222%	158%	176%	263%
S&P 500	57%	61%	107%	118%

Source: From *Firms of Endearment: How World-Class Companies Profit from Passion and Purpose*, 2nd edition, by Raj Sisodia, Jag Sheth and David Wolfe. Copyright © 2014 by Rajendra S. Sisodia, Jagdish N. Sheth, and the Estate of David Wolfe. Reprinted with permission of Pearson Education.

S&P 500 as well as Jim Collins's widely followed "good-to-great companies" over long periods, by as much as ten times in fact! If true, such discerning stock-picking powers would rank Professor Sisodia and his coauthors well ahead of the best equity investors of all time, including Peter Lynch, Sir John Templeton, and Warren Buffett. Moreover, Raj Sisodia would have the prized gift of being able to turn many common investors into millionaires, if not billionaires.

In determining which corporations merit their highly coveted designation of firm of endearment, Sisodia, Wolfe, and Sheth invite the readers of their work *Firms of Endearment* to reflect upon the words *affection, love, joy, authenticity, compassion*, and *soulfulness*. The authors then lay out their core definition of a firm of endearment:

> Quite simply, a Firm of Endearment is a company that endears itself to stakeholders by bringing the interests of all stakeholder groups into strategic alignment. No stakeholder group benefits at the expense of any other stakeholder group; each prospers as the others do. These companies meet the functional and psychological needs of their stakeholders in ways that delight them and engender affection for and loyalty to the company.

To put a fine point on this description, the authors say that firms of endearment never strive for a share of the wallet; instead, they strive for a share of heart. Their core values are also differentiating. They typically result in much more modest executive pay, very low employee turnover, customers fully satisfied by *every* experience, meager marketing costs, and a healthy obsession with creating a positive corporate culture. Sisodia and his colleagues also coined a catchy acronym, SPICE, so that readers more fully appreciate their "secret sauce." It turns out that firms of endearment care deeply about *s*ociety and about their *p*artners, *i*nvestors, *c*ustomers, and *e*mployees. If you are at all like me, all this may be creating a little heartburn!

So, who exactly, you must be asking, has achieved all these great, heartfelt, and optimally spicy objectives? It turns out that firms of endearment are not just U.S. publicly listed firms. They also include U.S. private corporations and more than a dozen non-U.S. companies. Some—like Southwest Airlines, L.L.Bean, the Motley Fool, and Toms—arguably fit the bill. These firms have established reputations for treating their customers, employees, and the environment in uniquely mindful and passionate ways. BMW, Honda, Toyota, Schlumberger, and the Korean steelmaker POSCO are also on the list, however, which seems somewhat incongruous. BMW has been embroiled in a series of class-action lawsuits for defects in their cars from 2013 to 2018, some of which may involve cover-ups. As a leading oil and gas servicing corporation, Schlumberger found itself badly buffeted by much lower development investment by the oil majors; between June 2014 and March 2020, their stock fell by 90 percent. Another private firm of endearment, Interstate Batteries, has publicly declared to "glorify God as we supply our customers." Its website quotes scripture as the basis of its unwavering commitment to operate "with a servant's heart" (1 Peter 4:10: "Each of you should use whatever gift you have received to serve others, as faithful stewards of God's grace in its various forms."). Its nearly unblemished record of creating and distributing hundreds of millions of batteries for cars, mobile devices, and household and medical equipment is undoubtedly laudable. The fact that its underlying business involves lithium and lead batteries—interests that may make some environmental activists less than enthusiastic—may give others pause, however. This is not a swipe at Interstate Batteries but rather a recognition that ESG risks may ultimately hit the company hard owing to material "E" concerns.

In 2019, Raj Sisodia joined forces with the organizational development genius Michael J. Gelb to coauthor *The Healing Organization*. Tom Peters, the acclaimed coauthor of the 1982 classic *In Search of Excellence*, which sold more than three million copies, states unequivocally in his foreword that Sisodia and Gelb's text on "awakening the conscience of business to help save the world" will appear on lists of all-time best business books. Why? Because it discusses many cases of corporations that fundamentally transformed themselves or their industries in this way.

According to Sisodia and Gelb, healing organizations, first and foremost, are "people-centric." They frame everything they do in terms of their direct impacts on the lives of people. They also adopt the same three core principles:

1) Assume moral responsibility to prevent and alleviate unnecessary suffering.
2) Recognize that employees are your first stakeholders.
3) Define, communicate, and live by a healing purpose.

At the top of a healing organization, moreover, healing *leaders* must preside. Take Jim Sinegal, the founder and former CEO of Costco, as an example. Sinegal intentionally paid his employees twice his direct competitors' wages while refusing to charge more than a 14 percent markup on anything his stores sold. This has meant that Costco has bred uncommon employee loyalty while selling many generic drugs at 90 percent less than you might find at Target, Walmart, or CVS. To be a healing leader, Sisodia and Gelb write convincingly, you must approach your business opportunity the way Sinegal approached his: with genuine concern for all of one's stakeholders. Business leaders should also embody the ten most essential character traits for long-term profit maximization:

1. Embrace innocence and humility.
2. Alchemize your suffering.
3. Be true to yourself.
4. Model the values and behaviors you want your organization to have.
5. Think creatively/lead innovation.
6. Be an inspiring storyteller.
7. Remember means and ends are inseparable.
8. Harmonize masculine/feminine and child/elder energies.
9. Think seven generations ahead—at least.
10. Always operate from love.

I am obviously no fan of Professor Sisodia's claim that he has been uniquely able to pick great stocks by examining a corporation's soul. That said, I do believe many corporations have achieved far more than others dreamed possible by explicitly embracing heartfelt purpose. I also admire empathetic leadership, for which Sisodia is an especially eloquent spokesperson. Most importantly, I believe the tools that Professor Sisodia, Just Capital, and B Lab are all separately looking to refine and deploy could be useful for maximizing the roles public and private corporations can and should play in promoting more inclusive, sustainable growth. They are together, of course, quite different from the paths regulators and accounting standards boards are fixated upon. ESG methodologies focus on materiality and consistency, not the harmonization of male and female energies or the importance of soulfulness and joy. I accept that these traits could well have a significant impact on corporate profitability and shareholder returns. We all should. However, they are simultaneously unlikely to displace the entire ESG movement as the primary ongoing force for reallocating global capital with the dual promise of social progress and superior returns. For this reason, we will need to return our focus back to the extant ESG movement and how it could be reconfigured for more meaningful and measurable impact.

We have spent this chapter investigating whether there is a superior, sys-temically meaningful, and scalable method for hardwiring goodness into corporate behavior beyond in situ ESG efforts. While we've seen some prom-ising approaches, the answer is no—no, there is not one all-encompassing method for recalibrating the corporate world into a more benevolent force that consistently optimizes the competing interests of all stakeholders. Nor is there an underused strategy that can somehow lead the corporate world to solve all of our environmental and social problems at scale. There are several that bear watching, however, and some that show significant potential for greater scalability. Meanwhile, ongoing ESG efforts across the entire corpo-rate, regulatory, and financial ecosystem, including serious efforts to develop a common lexicon, are and will likely remain the standard by which all other initiatives to hardwire corporate goodness are measured. In global finance, ESG is still the predominant force solving for double bottom lines. A lot is resting on its success and failure.

At the end of the day, asset owners are the ones who must decide how they will allocate their capital to generate the outcomes they seek. In such an environment, ESG customization is sure to become more critical. Some will favor financial returns over ethical considerations, others the converse. Still more will want their investments to generate great returns *and* have benefi-cial social and environmental outcomes. Investments that genuinely "do well and do good" occupy a rare sweet spot.

With all that's at stake, it is no surprise hundreds of firms are claiming they have a unique process or unmatched capability that enables them to generate superior returns and beneficial social outcomes. Given trillions of dollars are involved, it's also little wonder the financial services industry finds itself deeply enmeshed in an all-out ESG arms race.

Chapter Eleven

INSIDE THE ESG ARMS RACE

Winning isn't everything. It's second to breathing.
GEORGE STEINBRENNER

Wherever and whenever a great deal of money is in motion, great battles are being fought. Since 2018, an average of $8 billion has been moved from traditional investment strategies into ESG-themed products *every day*. A Bloomberg Intelligence report claims that more than one-third of all assets managed globally will carry an explicit ESG label by 2025, amounting by then to more than $50 trillion in total.[1] As rapidly increasing as recent ESG investment inflows have been, they seem destined to accelerate even further. Wherever great battles are fought, victors and vanquished emerge. Such is the case in the asset management and investment advisory business today. Thousands of asset management and investment advisory firms proclaim that their proprietary investment processes involving environmental, social, and governance analytics are uniquely suited to generate superior outcomes. But we all know this cannot be true: by definition, outperformance can be achieved only by a select few. The average return of all investment strategies will be . . . average.

Welcome to the ESG arms race.

One of the most significant battles playing out now is what genuinely qualifies as a "sustainable" ESG product. To some extent, the ESG market has been a classic case of eating one's dessert first and saving one's vegetables for last. Under the UN's Principles for Responsible Investment (PRI), firms need only promise to "integrate" ESG considerations into their investment practices. More than seven thousand institutions say they have done so. These firms collectively manage more than $120 trillion in assets. Their self-designated "integrated" ESG investment strategies include a wide array of products and methodologies spanning traditional screened indexed products, active thematic products, tilted-screened funds, impact strategies, and "double-bottom-line" variants. ESG-integrated investments also cover every asset class, from cash, public equity, fixed income, and loans to private equity, real assets, and commodities. Yet, as a group, asset managers and their overseers have done very little to classify ESG-integrated products transparently, consistently, or usefully for their customers. Instead, each has gone their own way. In some sense, they have all had their dessert first.

Predictably, the PRI's minimal accreditation standards have enabled a profusion of managers to say, "ESG is at the heart of everything we do." One firm making this claim—DWS Group, the asset management arm of Deutsche Bank—suffered a high-profile public embarrassment over the gap between what was said and what was being done. DWS's once-sustainability chief Desiree Fixler says she was fired for speaking up about the shortcomings in DWS's ESG integration procedures. "Posturing with big statements on climate action and inclusion without the goods to back it up is really quite harmful," Fixler told the *Wall Street Journal*.[2] An internal review of DWS's investment process allegedly found that only a "small fraction of the investment platform" properly integrated ESG criteria. While many of the PRI's seven thousand so-called ESG integrators are no doubt diligent in their procedures, some of the lesser players surely cut corners. In short, it has been all too easy for some firms to say they use ESG "considerations" in their traditional investment products without meaningfully changing any of their traditional practices. Getting to more rigorous standards involving accurate and actionable definitions about how companies' investment processes are genuinely sustainable—the eating-vegetables part—is very much the heart of the battle that lies ahead.

An example of the type of arm-to-arm combat now taking place broke out into the open after another recent Bloomberg Intelligence report.[3] Bloomberg's heads of ESG thematic investing accused BlackRock and their largest ESG ETF (ticker symbol "ESGU")—then the largest ESG ETF in the world, with $23 billion in assets—of inconsistent stock and sector allocation. At the time of the report's writing, ESGU owned oil and gas stocks worth 2.72 percent of the total fund's allocations—very slightly more than the S&P 500's weighting of 2.63 percent, as well as the 2.44 percent weighting of the MSCI USA Index upon which the strategy was based. Small holdings in

Honeywell, Huntington Ingalls, and Jacobs Engineering were also cited for possible infractions, as each has peripheral involvement in weapons production. Bloomberg claimed these overweights and shareholdings "cast a shadow" on both the MSCI index used, as well as the general claim that ESGU supports sustainable outcomes. Unfortunately, they failed to mention that the portfolio managers of ESGU were well within their pledged tracking error limits and investment guidelines. Perhaps even more critical, Bloomberg itself has a series of ESG indices that are vigorously competing to displace MSCI's. Wars are often fought in this way; their first casualty is the truth. If Bloomberg was intent upon full and fair disclosure, they would have acknowledged that ESGU was not being mismanaged versus its prospectus. They would have further observed that ESGU had outperformed both the S&P 500 and its chosen MSCI index since its inception, while broadly maintaining much less carbon intensity. They would also have disclosed that they may have been conflicted by their allegiance to competing processes and products. Finally, they would have mentioned that their indices also permit oil and gas holdings and controversial companies like Honeywell, hardly any different from MSCI's approach. If they didn't, they would risk serious tracking error versus the indices they claim they are trying to beat. But, hey, war is war!

In a 2021 report wryly titled *It's Not Easy Being Green*, the Deloitte-owned Casey Quirk details how hard managing authentic transformation from current unregulated ESG practices to verifiably sustainable investment practices will be for most firms (figure 11.1). Nearly half of the massive growth in ESG assets they foresee is expected to come from converting existing strategies to more verifiably sustainable and socially conscious funds. What exactly will change in these strategies? How will "greater sustainability" be verifiably measured and implemented?

In the end, Casey Quirk believes that ESG strategies will consist of four broad types: (1) *ESG risk integrators* who will use ESG data as one of many portfolio inputs, with few defined environmental, social, or governance outcomes explicitly promised; (2) *client-led ESG solutions* in which strategies that seek more dedicated social and environmental outcomes will be determined primarily through direct consultation with and customization for clients; (3) *goal-oriented ESG outcome providers* whose enterprise and investment decisions will be defined by a clear set of ESG objectives, most related to climate; and (4) *sustainability purists* who will claim explicit objectives such as temperature alignment or zero carbon emissions as their core competitive identities. According to Casey Quirk, each of these four

Dedicated sustainable investing archetypes: Priorities and challenges

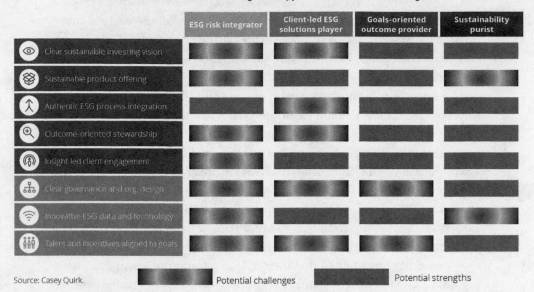

	ESG risk integrator	Client-led ESG solutions player	Goals-oriented outcome provider	Sustainability purist
Clear sustainable investing vision				
Sustainable product offering				
Authentic ESG process integration				
Outcome-oriented stewardship				
Insight-led client engagement				
Clear governance and org. design				
Innovative ESG data and technology				
Talent and incentives aligned to goals				

Source: Casey Quirk.   Potential challenges   Potential strengths

FIGURE 11.1. **"Winning the ESG arms race will require different competi-tive advantages relative to specific sectors of the sustainable investment market..."—Casey Quirk, 2021.**

Reprinted with permission from *It's not easy being green: Managing authentic transformation within sustainable investing.* Copyright © 2021 Deloitte Development LLC. All rights reserved.

strategic categories will also experience diverging growth dynamics, with ESG integrators—the most common now— falling from 49 percent to 34 percent of the overall market. They expect client-led solutions to decline in relative terms as well, from 43 percent to 25 percent. Goal-oriented products are expected to rise nearly fivefold, from 8 percent to 39 percent of the mar-ket, whereas sustainability purists will hold steady at 1 to 2 percent, thus remaining niche strategies.[4] Rapidly growing goal-oriented products may be broadly benchmarked against index-related returns and include additional specifics relating to multiple environmental or social objectives. They may also pursue impact outcomes in a more explicit manner (foreshadowing my soon-to-be-unveiled 1.6 percent "solution"). Asset managers and investment advisers will need to be transparent about what they can and cannot provide across each of these four strategies in truly differentiating ways. If not, they will lose the ESG market-share war.

One firm ready and willing to fight for more client-designed ESG wins is Dimensional Fund Advisors. In a 2021 report, they argue that the subjectivity, opacity, and complexity of ESG ratings have limited usefulness for investors, the same conclusion we arrived at in our discussion of materiality.[5] Instead, they claim a better approach is for investors to determine their ESG priorities, identify data by which those priorities may be verifiably measured, and integrate those goals into a comprehensive, risk-aware investment framework: "Rather than rely on generic ESG ratings, we believe investors would be better served to identify which specific ESG issues are most important to them, understand the data used by their investment manager . . . and ask to see transparent reporting on specific ESG outcomes." Dimensional Fund Advisors expects their client-customized ESG strategies will largely focus on climate change, as they believe this is the primary sustainability challenge of our time.

Deutsche Bank's asset management and ESG research teams, headed by Brian Bedell and Debbie Jones, similarly project accelerating organic growth for "sustainable" investment products in the years ahead. They expect the battle to win trillions of ESG dollars in motion will play out across two broad fronts: *active* and *indexed* strategies. They note that the recent COVID-19 pandemic–related outperformance of many ESG-themed products is helping to fuel growing retail demand in the short run. So are rising concerns about climate change, the growing impact of European regulation, and the broader use of artificial intelligence programs. Because of these factors, Bedell and Jones expect ESG "targeted and thematic ETFs will be especially appealing to retail investors," driving more flows into indexed ESG products like ETFs.[6] Of course, the ESG ETF market has already been picking up steam. Between 2018 and July 2021, the total number of ESG ETFs listed globally nearly tripled, from 285 to 770. Total assets managed in ESG ETFs over the same period also grew by more than ten times. As active managers "strengthen their sustainable investing capabilities, including selecting securities with changing ESG characteristics," Bedell and Jones see an eventual rebound in active versus indexed strategies.

Reviewing several examples of indexed and active ESG strategies involving equity and fixed-income markets will provide a helpful foretaste of what's to come.

I. INDEXED EQUITY ESG STRATEGIES

One of the fastest-growing indexed equity ESG strategies screens out firms based upon a single variable: "temperature alignment." Asset managers and index providers are spending a great deal of time and effort identifying companies that have taken steps to become verifiably aligned with the Paris target of net zero carbon emissions by 2050. However, there is a significant discrepancy between index providers and individual analysts about which companies are and are not transition ready. For example, as of August 2021, analysts at S&P concluded that 305 members of the S&P 500 were eligible for inclusion in their Net Zero 2050 Paris-Aligned ESG Index; by inference, 195 were not.[7] To be included, an S&P 500 company has to provide evidence that they will get their emissions to zero by 2050 on a net basis using Trucost Transition Pathway analytics. They must also not be involved in controversial weapons; tobacco; exploration for coal, oil, or natural gas; or any other greenhouse gas–emitting industry that S&P considers disqualifying. In their analysis of all five hundred companies, S&P set and measured their own alignment rules. Other index providers have different methodologies and conclusions.

Amundi's Lyxor Asset Management arm has adopted S&P's index in several of their ETFs, taking S&P's 305 aligned companies from theory to practice. Lyxor is one of Europe's oldest and largest ETF providers. According to Bloomberg data, during the first fourteen months of trading, Lyxor's U.S. dollar– and euro-denominated ETFs that sought to replicate the S&P Net Zero 2050 Paris-Aligned ESG Index underperformed the index by about 250 basis points, or 2.5 percent. This is an unusually high amount of negative tracking error for an ETF. Lyxor also then charged 70 basis points of fees for these products, expensive by industry standards. However, through September 2021, after fees, Lyxor's S&P 500–screened net-zero ETFs still outperformed the unscreened S&P 500 by 77 basis points. In other words, S&P's screened index with 305 stocks outperformed the broader 500 index sufficiently to more than make up for Lyxor's high fees and tracking error. This means that S&P's temperature-aligned benchmark generated alpha relative to the broader S&P 500 over the period measured. Of course, owning S&P's temperature-aligned index outright would have been better, but Lyxor's S&P 500 temperature-screened ETF still produced some excess return.

Structuring temperature-aligned funds is challenging because they rely upon corporate data that remains frustratingly incomplete, as well as analytics that may differ markedly from firm to firm. Digging deeper provides perspective on the challenges that lie ahead for temperature-aligned products. For example, the think tank InfluenceMap recently analyzed 723 equity funds that purported to be aligned to Paris objectives in some manner. Their analysis found that funds that claimed to be compliant had no greater temperature alignment than did broad MSCI world indices or similarly broad indices with a few fossil fuel exclusions. Specifically, their research highlighted "a lack of consistency and often poor transparency on the alignment of many ESG and climate-themed funds with global climate targets."[8] Their report was intended to call out and help mitigate widespread greenwashing, a common practice wherein firms and funds pretend to have stronger environmental credentials than is actually the case.

Over time, it remains to be seen whether indices composed of companies already aligned with Paris will outperform broader indices that retain companies and industries that have yet to be. However, there is at least one intuitive reason to think they won't. Financially, it will always be preferable to own companies *before* they adopt best practices. Investors benefit most through the additional value created during and after positive transformation. This is the precise point the Norwegian central bank made in rejecting the adoption of ESG or climate-aligned indices: "Any decision to replace the current equity index with a climate-adjusted index would mean the fund misses out on opportunities where companies not included in the index undergo a transformation." At the same time, if more and more investors adopt climate- or otherwise ESG-screened indices, this would likely create greater demand for the same names. Ceteris paribus, rising demand should drive prices for the debt and equity instruments of transformed companies higher. If trillions of dollars have yet to be moved into products with verified ESG credentials, a self-fulfilling virtuous circle of higher prices for temperature-aligned companies becomes more likely. It is simply not possible to know for sure today whether temperature-aligned funds will outperform broader indices, though. What we can say with certainty is that anyone promising that such funds will definitely outperform broad indices while simultaneously creating a greener planet is claiming more than facts and simple common sense allow.[9] The cost-of-capital transmission mechanism upon which this argument broadly depends has not yet reduced greenhouse gas emissions or muted climate change in any verifiable manner. Nor does correlation establish causation.

II. ACTIVE EQUITY ESG STRATEGIES

Active strategies not only permit more tracking error versus an index, they explicitly encourage it. Their expressed purpose is to outperform an index. The best-performing active strategies do so by taking concentrated stock bets. One chooses an active manager over an index manager because one has a high degree of confidence in that manager's skill at stock selection as well as greater tolerance for volatility. A tolerance for volatility means one is willing to accept unexpected price moves beyond what a stated index suggests. Active managers can go through periods of underperformance—in fact, it would be odd if they didn't. If they do their jobs well, however, they can also be the one out of seven who consistently outperforms the market index they are trying to beat. As of this writing, only about four hundred actively managed ESG-themed equity funds have a track record of three years or longer. One of the best performers in this category—consistently ranking in the top decile versus their peers over three- and five-year horizons—is the Large-Cap Sustainable Growth fund managed by Karina Funk and David Powell of Brown Advisory.

Though measured against the Russell 1000 Growth Index, Funk and Powell typically own no more than thirty to forty names. To generate alpha versus their index, they heavily overweight companies with compelling valuations and sustainable competitive advantages. In the three years leading up to June 2021, Brown Advisory's Large-Cap Sustainable Growth Strategy added more than 225 basis points of outperformance per year. Given they charge an average fee of about 54 basis points, as long as higher volatility is not a problem, investing with Brown Advisory's sustainable growth team could be more attractive than owning the Russell 1000 Growth index outright. "Funk has a long history of conducting environmental, social, and governance analysis," Morningstar reported in assigning the fund one of their top ratings. "They seek companies with durable ESG-driven competitive advantages that ultimately lead to cost reduction, increased revenues, and enhanced franchise values."

Another emerging class of active ESG equity strategies that bears watching involves taking significant positions in companies with persistent environmental and social problems and engaging them to change. In nearly all cases, this form of engagement will not be welcomed. Fighting for board seats—as Engine No. 1 successfully did with Exxon—may become a common approach for targeted companies in such strategies. However, these are high-risk endeavors. Success largely entails displacing existing managements and reconfiguring long-held business practices. This is always hard, and failure would be quite costly. A final example of active ESG equity strategies worth considering explicitly targets measurable impact relative to environmentally or socially desirable outcomes like the UN's Sustainable Development Goals.

These strategies seek to "do well and do good" in a verifiable manner. I will discuss a wide range of these types of impact strategies in chapter 16.

III. INDEXED FIXED-INCOME ESG STRATEGIES

Turning from ESG equity strategies to ESG fixed-income strategies, one finds very different return prospects, in both nominal and real terms—that is to say, both before and after inflation is considered. For example, while many equity strategies have generated double-digit returns in recent years, most strategies involving high-quality fixed income have produced very low single-digit returns. Low fixed-income returns are the unavoidable consequence of maintaining highly accommodative monetary policies over long periods. To combat economic weaknesses and uncertainties caused by the COVID-19 pandemic, central banks in many countries reduced their official rates to zero—and in the cases of countries like Denmark, Japan, Sweden, and Switzerland, even *below* zero. The yields of most high-quality debt instruments have axiomatically followed suit. For most of the past few years, to have a yield higher than 1 percent, one had either to assume significant credit risk—meaning more uncertainty about the borrower's ability to repay you— or more duration risk. *Duration* refers to the time that transpires between a security's purchase and when its full cost is returned. Today, getting a yield of more than 2 percent while taking minimal credit risk may mean you have to wait as long as ten, fifty, or even one hundred years before your initial capital is returned. While fixed-income securities do not provide much yield, they can add a form of resilience to a portfolio. More often than not, when equity prices fall sharply, bond prices rise. Portfolios with balanced combinations of uncorrelated assets typically incur less overall volatility.

We have not yet spoken about green bonds, or their cousins, social and sustainability bonds. As these instruments show considerable promise for promoting inclusive, sustainable growth, it's important we speak more about them.

A green bond is a fixed-income instrument, the proceeds of which support specified climate-related or environmental outcomes. They often entail the same credit risk as their issuer's general obligations but may be more highly rated if they are asset backed and lower rated if they are not as senior in the capital stack. Social bonds are similar, except their proceeds support projects of a nonenvironmental nature, like more affordable housing or broader access to education. Unlike ESG equity strategies, which largely rely upon the cost-of-capital transmission mechanism to generate impact, green bonds and their social counterparts can and usually do result in specific, demonstrable change. The proceeds from every green and social bond are almost always earmarked for identified pledged outcomes that can be verified. These results can be meaningful. For example, NN Investment Partners in the Netherlands reported during 2021 that their holdings of climate-related green bonds added 333 megawatts of renewable energy capacity and 835 gigawatt hours of annual renewable energy output, thereby negating 561,211 metric tons of CO_2 emissions. These are laudable outcomes, wholly consonant with our goal of more inclusive, sustainable growth. NN also states that $1 million invested in their traditional green bond fund is the environmental equivalent of powering twenty-five EU households for a year while simultaneously removing 165 passenger cars from the road.[10] This means that investments with NN partners are helping the planet become more resilient while generating market-based returns. Green bonds also now rank among the fastest-growing debt markets in the world. According to the Climate Bonds Initiative, the annual issuance of green bonds could exceed $1 trillion by 2023 (figure 11.2). Cumulative issuance since 2015 is nearly $2 trillion. In sum, green, social, and sustainability-linked debt instruments rank among the most promising fixed-income products that are verifiably doing good while still producing some return. Their principal drawback is that those returns are far below

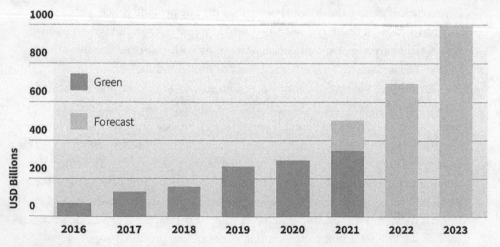

FIGURE 11.2. Green bonds are debt instruments that verifiably do "good." They are also one of the fastest-growing markets globally.
© Climate Bonds Initiative, 2021.

the levels equity markets can and have been generating, and will likely prove insufficient on their own to promote pension plan solvency.

One example of an indexed ESG fixed-income product is the VanEck Green Bond ETF. Operating under the ticker symbol "GRNB," it seeks to replicate as closely as possible, before fees and expenses, the price and yield performance of the S&P Green Bond U.S. Dollar Select Index. According to Bloomberg data, it has fallen somewhat short of doing so; since its inception in March 2017 through December 2021, GRNB returned about 3.3 percent per annum versus its index of 3.6 percent per annum. Green bonds and their social cousins certainly get high marks for promoting inclusive, sustainable growth. They are clear examples of the types of instruments that can help fashion the good society. But it must be simultaneously said their target return rates are very modest. Given this, taken on their own, they cannot serve as sufficient building blocks for one's long-term retirement portfolio. If one owns bonds, however, a good case could be made for a meaningful green or other sustainable bond allocation. This suggestion will be part of my proposed 1.6 percent "solution," soon to come.

IV. ACTIVE FIXED-INCOME ESG STRATEGIES

One final example of an investment strategy in which the ESG arms race is playing out involves active fixed income. As with active equity strategies, active fixed-income strategies seek higher returns than an index through more concentrated exposures driven by intensive research. Target returns are still below the equity markets because the eligible universe of investments is low-yielding. In the prevailing rate environment, one must assume large amounts of duration and credit risk to get returns well above the rate of inflation. With active management, higher returns are possible. Of course, lower returns are possible, too!

One recent example of an active fixed-income ESG strategy that may appeal to a growing number of investors seeking to do both well and good can be derived from BlackRock's proprietary fixed-income framework, dubbed "PEXT/NEXT." This broad strategy seeks to overweight debt instruments that generate *positive* externalities while eliminating exposures that may cause *negative* externalities across all industries—hence *PEXT/NEXT*. Green bonds, social bonds, and electric vehicle asset-backed securities are examples of the former. Predatory lending, tobacco, gambling, oil sands, coal, and adult entertainment are examples of the latter. Of course, most fixed-income securities fall between these extremes; in BlackRock's termi-nology, this includes so-called baseline externalities, or "BEXT" bonds, and "discussion bonds," or "DEXT" bonds. Portfolio managers using the PEXT/NEXT methodology must avoid all bonds in the NEXT category while exer-cising proper discretion in the other three categories.[11]

BlackRock's PEXT/NEXT framework is new, meaning there is not yet a multiyear track record to back up its potential for the future. Still, there is reason to believe PEXT/NEXT could mirror the historic performance of BlackRock's active fixed-income team. It will be based upon their historic methodologies and powered by BlackRock's proprietary risk analytic sys-tem, Aladdin. In short, PEXT/NEXT has a reasonable chance of generating returns above those of the indices it pledges to follow while simultaneously promoting certain desirable social and environmental objectives in a verifi-able manner. PEXT/NEXT is an active fixed-income framework that could do well and do good.

These four examples of ESG strategies—one each for active and indexed strategies in equity and fixed-income markets—illustrate a small fraction of the thousands of investment products that hope and expect to be included in the accredited "sustainable" ESG taxonomies of the future. Ultimately, success in the ESG arms race will require successful positioning vis-à-vis emerging consumer desires, competitive pricing, and evolving regulatory requirements. Concerted industry input will likely impact the latter as well. Regulatory decisions are seldom taken in a vacuum. Those that are often do more harm than good.

To this end, my "dessert before vegetables" analogy may have another practical application. It relates to the unfolding dynamic of consumer pref-erences, industry positioning, regulatory practice, and the ultimate process of delineating ESG products into more coherent taxonomies.

In 1990, the U.S. Congress passed the Organic Foods Production Act. The purpose of the law was to establish widely agreed and accepted standards for farming and labeling organic food. The act was largely a response to multiple petitions and spirited complaints from established organic industry groups who believed the complexity and inconsistency of prevailing reporting stan-dards were causing customer confusion and undermining confidence in the integrity of organic products. Nonconformity was also harming the pock-etbooks of established organic producers. With no barriers to entry, grow-ing numbers of farmers claimed their products were as organic as any other without taking any special measures, such as properly curing their fields, using qualified seeds, or organically fertilizing their crops. Such shortcuts gave new entrants a cost advantage over more labor-intensive and complicated procedures. Established organic food artisans sought greater regulation to protect consumers from organic quacks; at the same time, more stringent regulations were essential to their survival.

The 1990 act established a fifteen-member National Organic Standards Board to develop standards that the organic farming community could broadly accept and maintain. As it so happens, their work wasn't easy. It also wasn't swift. It took the board seven years to agree upon organic standards and draft proposed rules. And that initial proposal drew so many responses that it was another three years, in 2000, before the final rules were ultimately codified. In sum, one decade of continuous work with relevant and vested parties was needed before accredited U.S. Department of Agriculture "organic food" seals could be placed on fruits and vegetables. Until that time, unregulated, producer-only labels claimed the same thing. In short, after ten years of concerted disagreement, debate, and deliberation, "organic" labels on fruits and vegetables finally meant something more than what their producers claimed on their own.

I believe this ten-year codification process for organic food serves as a helpful benchmark for the process of determining sustainable investing taxonomies for the future. Under the direction of some central regulating bodies, a cross section of asset management firms, asset owners, and other financial market professionals needs to debate and ultimately agree upon widespread, uniform principles and definitions to protect consumer interests and properly codify sustainable products. But even this is much easier said than done. While "sustainable investments" are often global, in the end they must all be listed and regulated locally. As such, they are subject to the considerations of multiple overlapping regulatory jurisdictions and competing interests. Rather than one central regulating body, there will be many. How much they coordinate with one another remains an open question. A fair assumption is not nearly enough.

The European Union has taken the lead in defining sustainability through its Sustainable Finance Disclosure Regulation (SFDR). The SFDR "lays down harmonized rules for financial market participants and financial advisers on transparency with regard to the integration of sustainability risks and the consideration of adverse sustainability impacts in their processes and the provision of sustainability-related information with respect to financial products." After three years of industry comment and preparation, the SFDR took effect on March 10, 2021. The primary purpose of the regulation was to increase transparency about the integration of sustainability risks in the financial sector. It requires disclosures of both positive and negative externality effects. But the diminution of greenwashing and more consistent product taxonomies are not the only things the European Union is seeking. In addition to fostering greater transparency, European regulators hope to reorient capital flows toward more sustainable growth methods and preferred social outcomes, like directly muting climate change risks, promoting gender equity, and reducing pay gaps. Today, financial products marketed in the European Union have to be designated one of three types: (1) Article 6 funds, which encompass all generic fund products; (2) Article 8 funds, which claim to promote some environmental or social characteristics; or (3) Article 9 funds, which have an explicit sustainable investment objective, such as temperature alignment. Most importantly, firms that designate their funds as either Article 8 or 9 must simultaneously provide a clear and reasoned explanation of how their generic or specific sustainability objectives will be achieved.

Morningstar analyzed the early implications of this SFDR rule, reviewing 5,695 funds domiciled in Luxembourg, Europe's most prominent fund locale. They also surveyed thirty large asset managers, compiling their self-defined lists of Article 8 and 9 funds. As asset managers took different approaches to classifying their funds, only about 25 percent of total European fund assets ended up as "sustainable"—one in four—versus nearly 80 percent that had claimed to be "ESG integrated" before the SFDR took effect.[12] Morningstar expects this 25 percent total will rise over time as managers enhance their existing strategies, reclassify their existing funds, or launch new ones specifically designed to fulfill Article 8 and 9 strictures. Unsurprisingly, French, Dutch, and German firms jumped out in front of the pack for market share.

While the European Union moved first to set SFDR classifications, and European asset managers have taken an early lead in the battle for sustainable fund market share on their home turf, some U.S. rules and regulations are not far behind (figure 11.3). "Funds that say they're sustainable and green and the like—what stands beneath that right now?" the SEC chair, Gary Gensler, asked EU lawmakers rhetorically on September 1, 2021. Gensler has not been shy about his intentions: "When it comes to sustainability-related investing, there's currently a huge range of what asset

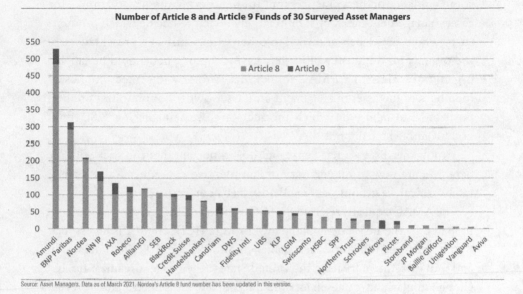

Number of Article 8 and Article 9 Funds of 30 Surveyed Asset Managers

Source: Asset Managers. Data as of March 2021. Nordea's Article 8 fund number has been updated in this version.

FIGURE 11.3. French, Dutch, and German firms lead the early race for EU-defined sustainable funds.
From "SFDR – The First 20 Days: What the early batch of new disclosures are telling us so far," by Hortense Bioy, Elizabeth Stuart and Andy Pettit. Morningstar, 30 March 2021. © 2021 Morningstar, Inc. All Rights Reserved. Reproduced with permission.

managers might mean by certain terms or what criteria they use. I think investors should be able to drill down to see what's under the hood of these funds." That said, it very much appears the SEC is likely to take a principles-based approach to classification rather than the harder product-based approach favored by the European Union. The SEC's Asset Management Advisory Committee (AMAC)—a group of appointed representatives from the industry who provide recommendations to the SEC—strongly recommends that the SEC not stifle innovation. "The AMAC recommends the SEC take steps to foster meaningful, consistent, and comparable disclosure of material environmental, social, and governance (ESG) matters by issuers."[13] In other words, the AMAC's top priority is to get additional hard data from corporations. In the interim, "the SEC should suggest best practices to enhance ESG investment product disclosure, including alignment with the terminology developed by the Investment Company Institute ('ICI') ESG Working Group." Translation: We promise to do our part, but this process will take time. Yes, the SEC can and should offer some guidance on best practices, but in the interim, the AMAC thinks the asset management industry should largely be left to decide which products carry which ESG labels and why. In other words, just as representatives of the organic food industry ultimately agreed upon what should be labeled organic over time, the AMAC thinks that similarly informed and vested members of the asset management and asset owner communities are best positioned to advise on which products are "sustainable" or "ESG integrated" and which are not.

I know this discussion is dense, but it is important that you stick with me here. We are now in the territory of "unintended consequences." It's absolutely critical we get our next regulatory steps correct.

The AMAC explicitly referred the SEC back to an ESG taxonomy document prepared by the Investment Company Institute (ICI). The ICI is an association of regulated funds and their managers and includes ETFs, mutual funds, and unit investment trusts in the United States. Their sister organization, ICI Global, similarly advocates in jurisdictions outside the United States. A recent ICI Global white paper titled *Funds' Use of ESG Integration and Sustainable Investing Strategies* outlines several broad definitions for "ESG integration" and "sustainable investing," which may well withstand the test of time:

With **ESG integration**, funds seek to enhance their financial performance by analyzing material ESG considerations along with other material risks.

With **sustainable investing**, funds use ESG analysis as a significant part of their investment thesis to meet investors' objectives while seeking financial returns.

Funds that pursue sustainable investing commonly do so with one or more of the following three approaches:

1) With **ESG exclusionary investing**, funds aim to exclude companies or sectors that fail to meet specific sustainability criteria or don't align with their investment objectives.

A fund taking this approach might opt against investing in companies that do business in weapons manufacturing or distribution, gambling, tobacco, alcohol, or nuclear energy.

2) With **ESG inclusionary investing**, funds generally seek positive sustainability-related outcomes by selecting investments that tilt a portfolio based on ESG factors alongside a financial return.

A fund taking this approach might invest in companies that benefit from clean energy generation, sustainable infrastructure, waste management, or other environmentally friendly activities.

3) With *impact investing*, funds seek to generate positive, measurable, and reportable social and environmental impact alongside a financial return.

A fund taking this approach might invest in bonds that finance affordable housing, health care, or mass transit.

By using this terminology, funds can communicate clearly and consistently about their ESG-related investment strategies—and help investors make more informed decisions.[14]

While I expect something like this broad framework will ultimately win the day, these definitions are as important for what they do not say as for what they do. Unlike pending regulations in the European Union, for example, ICI Global advocates bucketing sustainable investing using three broad *principles*: exclusionary, inclusionary, and impact. As you can see, these seem to hold "ESG integration" to no more exacting standard than traditional financial analysis. More likely than not, were they to be adopted, it would spell the end of all ESG integration strategies; in time, they would merely be subsumed back into traditional analysis. Frankly, this is exactly where they belong. Accounting for environmental, social, and governance risks is very important—but it should already be common practice at every investment management firm. After all, ESG risks are real. Risk analysis that does not systematically account for "E," "S," and "G" concerns in its regular course of business is flawed. Investment managers should not be operating with multiple risk systems—that is, with one that employs ESG analytics and another that does not. They should use one comprehensive framework in which ESG analysis is fundamentally incorporated. In this respect, it's probably time for generic ESG labels to go the way of the dinosaurs. Ubiquitous ESG labels aren't providing any additional clarity; instead, they are breeding obfuscation. Let's stop calling generic funds "ESG."

But there is yet another danger in becoming overly prescriptive with definitions and taxonomies, one that ICI Global has gone to great lengths to emphasize in Europe. "Crowding investors into product categories 'created' by regulation with a small investable universe does not further the objective of protecting investors," the chief counsel for ICI Global, Jennifer Choi, writes. "We caution that threshold criteria would likely narrow the existing diversity of sustainability-related strategies and reduce investor choice, without improving transparency or investors' understanding of sustainability-related strategies."[15] In other words, don't tip the scale toward a restricted opportunity set without expecting unsustainable price distortions to result.

ICI Global is not expressing concern about crowding investors into regulated product categories with restricted investment universes prematurely. By some professional estimates, we are already past the point of overly crowded ESG trades. We learned in the global financial crisis that officially sanctioned bubbles portend deeper troubles. We also experienced this in the savings and loan crisis of the early 1980s. Good intentions create unsustainable market valuations. When those valuations correct, hell can be unleashed.

Indeed, officially sanctioned or officially aided and abetted financial bubbles are almost always the worst kind. The investing public trust authorities will get it right. Unfortunately, this is very hard to do even when one's goals are genuinely laudable.

Chapter Twelve

CROWDED TRADES

Don't follow the crowd. Let the crowd follow you.
MARGARET THATCHER

Bull markets are born in pessimism, grow on skepticism, mature
on optimism—and die on euphoria.
SIR JOHN TEMPLETON

About one thing there should be no debate: the current ESG investment
phenomenon has all the makings of a potentially catastrophic investment
bubble. While many consider historic ESG valuations fair, based upon current dynamics, it's difficult to imagine how the markets will avoid episodic
ESG dislocations and painful sell-offs. Crowded trades have a way of becoming less crowded. When they do, it's seldom pleasant. Moreover, the implications of a systematic, ESG-led sell-off would be profound.

Several years from now—three, maybe fewer—I doubt there will be many
generic "ESG" strategies left. Why? I believe they all will have been renamed,
refitted, or reassumed under historically standard investment categories, like
factor strategies that target quality, growth, or minimum volatility. In fact,
this process of winnowing down strategies deemed "ESG" had already begun
at the end of 2021, when Morningstar dropped their "sustainable" label from
1,200 European-domiciled funds, equivalent to 27 percent of their entire
universe. It's clear many thematic, screened, and tilted funds will remain,
but calling them "ESG" won't provide any additional insight or clarity about
how they are managed. One important reason for this was mentioned at
the end of the last chapter: sooner or later (though preferably sooner) all
ESG risk analysis needs to be subsumed back into formal risk analysis. ESG
risk analysis should not be something supplemental or somehow added as a
bonus—the fancy china you roll out when your most important guests arrive
for dinner. Incorporating knowable environmental, social, and governance

risks in one's investment process should already be "standard operating procedure" at every investment firm. If your risk model does not already incorporate environmental, social, and governance concerns, your risk model is flawed. Fix it!

The second reason I think a shakeout or some form of reckoning in ESG-named funds may occur relates to performance. As currently composed, I expect many ESG funds will either generate more volatility versus market indices or less social benefit than implicitly presumed. This is merely a restatement of what we learned in our earlier discussion of values and valuations. Given most ESG strategies are broadly expected to deliver double-bottom-line results—superior returns *and* environmental or social progress—it's increasingly apparent a good number will not succeed on one or both counts. In the interim, more promising active strategies and the benefits of more explicit impact strategies like those we began to discuss in the last chapter (and will soon consider in greater detail) will have stepped forward and distinguished themselves. For example, green bonds and their social cousins are already proving to be salient tools for promoting a growing number of the UN's laudable Sustainable Development Goals. Some active ESG equity managers armed with transformational tactics like those of the Shared Value Initiative or other unique methodologies will also likely achieve notable double-bottom-line results. In short, winners will begin to separate themselves from losers. Indexed ESG strategies in particular will increasingly be seen as risk mitigation tools rather than products that verifiably promote more salient corporate behavior. This is not to say they won't succeed at risk mitigation. Some have and will likely continue to do so over time; they will just become more limited in scope.

Why do I make these statements with such firm personal conviction? On the bond side, central banks have intentionally engineered negative real rates to reduce excessive indebtedness. Besides more buoyant economic growth and painful defaults, negative real rates and higher inflation are the only effective tools policy-makers have to alleviate unsustainable debt burdens. Few fixed-income strategies will return anything close to the rate of inflation in the short term because that is what central banks now require. Those that do will need to deploy deft, active credit risk-taking as well as skillful market timing. All of this will be hard.

On the equity side, we established earlier the primary valuation challenge ESG funds face. A lot of money is chasing a limited number of corporations with verifiably pristine environmental, social, and governance credentials.

The valuations of many of these "good" ESG companies versus their less laudable peers became very extended by late 2021, unsustainably so in my view. It is possible they could get more expensive on a relative basis—but equally possible that some mean reversion will occur. In fact, the first quarter of 2022 has already shown how painful this type of mean reversion can be.

Concomitantly, unverified and unverifiable performance pledges are being exacerbated by well-intended—indeed, even necessary—regulatory efforts that aim to classify investment products more reliably as "sustainable." As these official classifications tighten, it is possible that many already crowded trades may end up becoming more crowded. As we've discussed, this is the axiomatic consequence of "good intentions"; the official designation of any investment as "sustainable" officially suggests that its value will be sustained. Official certifications like these lure investors who might otherwise be more skeptical and discerning to reposition their capital; why, after all, would any regulator encourage the public to buy something that was unsafe? But the prices of all securities and products—sustainable or not—are destined to incur volatility. When they do—not *if* but *when*—the most overvalued securities invariably fall the most. Why? Because over time, Benjamin Graham's voting machine becomes a weighing machine. Profits must eventually justify inflated price–earnings (PE) multiples, or those inflated multiples must come down. This is simple financial physics.

In 2013, Eugene Fama shared the Nobel Prize in Economic Sciences with Robert Shiller and Lars Peter Hansen for their work on asset pricing. The centerpiece of their research has become known as the "efficient market hypothesis." This hypothesis states that financial assets can and will be priced appropriately once their inherent properties are known.[1] It means that once all material information is revealed to the market, stock and bond valuations will be discounted appropriately. This is why the work of SASB, TCFD, and others is so important; they are trying to surface material risks that are not in plain sight. Like most time-tested hypotheses, the efficient market supposition is more true than untrue. That said, there are many factors that may cause asset prices to deviate from their inherent values in the shorter term. Some of those factors are cited by Sir John Templeton in the epigraph at the beginning of this chapter: emotions like pessimism, skepticism, optimism, and euphoria. As Sir John wryly notes, however, investor sentiments more often than not are reliable *counter*indicators. By consistently leaning against them, a handful of legendary investors like Sir John himself amassed multibillion-dollar fortunes.[2]

John Bogle put his own, typically direct spin on ESG's possible bubble predicament: "When counterproductive investor emotions are magnified by counterproductive fund industry promotions, little good is apt to result." To be clear, I don't consider the ESG movement to be counterproductive; I think society has and will continue to benefit from the promotion of more mindful business practices, something the ESG movement has brought to scale. But I worry that too many believe their current ESG allocations are destined to outperform while simultaneously making all our environmental and social problems go away. In fact, the investment performance of ESG strategies has been and will likely remain quite mixed. Meanwhile, we all know our environmental and social challenges require much more comprehensive solutions. The ESG movement has not been all we need and want it to be.

In my personal opinion, ESG investing as a broad investment adventure has mostly vacillated between optimism and euphoria during 2020 and 2021. The year 2022 seems destined for broad ESG underperformance. Longer term, I remain agnostic on ESG investment performance. Some strategies will work, others not. We'll just have to see.

As 2021 drew to a close, all equities and fixed-income assets were at historically high levels. Another helpful equity valuation tool invented by Robert Shiller—the cyclically adjusted price–earnings (CAPE) ratio—illustrates this convincingly. Shiller's CAPE ratio calculates the value of stock prices versus their historic, cyclically adjusted earnings. The higher the index trends, the more elevated stock prices are relative to their historic earnings capacity. At some point, CAPE ratios will revert to their historic mean. As figure 12.1 shows, the CAPE ratio of the S&P 500 as of late 2021 had been higher on only one other occasion, during the "dot-com frenzy" of 1999 and 2000. Within two years of that peak, stock investors were $5 trillion poorer. As the figure shows, stock valuations were even more extended in late 2021 than they were on Black Tuesday, back in 1929. Over less than forty-eight hours, on Black Monday and Tuesday in 1929, stocks lost one-quarter of their total value, the statistical equivalent of losing more than $15 trillion today. The Great Depression followed.

But even this CAPE ratio graph tells only part of the recent "ESG overvaluation" story. While all stocks were unsustainably expensive at the end of 2021, it turns out that "good" ESG stocks were even more so. Analysts at Bank of America Securities actively segment firms by their ESG ratings, comparing higher ranked or so-called good ESG companies with less highly ranked firms. They also compare good versus bad ESG firms relative to their non–cyclically adjusted PE multiples. Their analysis dating back to 2007 shows the

Shiller CAPE ratio

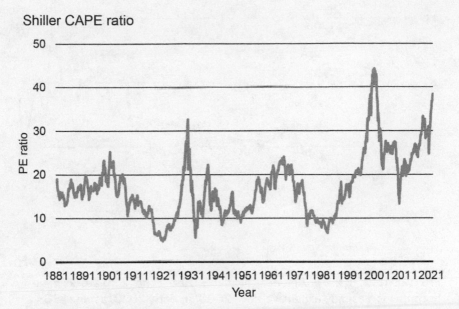

FIGURE 12.1. Equities were more overvalued in late 2021 than at any other time but 1999 and 2000—right before the dot-com bubble burst.
Figure by the author.

aggregate PE multiples of top quintile–versus–bottom quintile "good" and "bad" ESG firms have fluctuated between 85 percent and 155 percent. This means that the PEs of the top fifth of ESG firms have traded as low as a 15 percent discount to the bottom fifth and as high as a 55 percent premium.[3]

Figure 12.2 represents this trend with an upward-sloping line. In effect, this shows how "good" ESG firms have been steadily gaining in value versus "bad" ESG firms for years. It shows how and why many ESG strategies generated some historic outperformance. But can this trend continue forever?

Fifty years ago, a member of the Council of Economic Advisers and noted University of Chicago economics professor, Herbert Stein, made a declarative statement that has since come to be known as Stein's law: "If something cannot go on forever, it will stop." Self-evidently, the trend toward ever more expensive ESG stocks cannot go on forever. No expensive stock becomes more expensive *in perpetuity*. Figure 12.2 also shows that in early 2020, there was a significant revaluation of "good" ESG firms—from an all-time high of a 55 percent premium back to near parity. This means that good, sustainable, highly ranked ESG firms lost a lot of their premium in a very short period

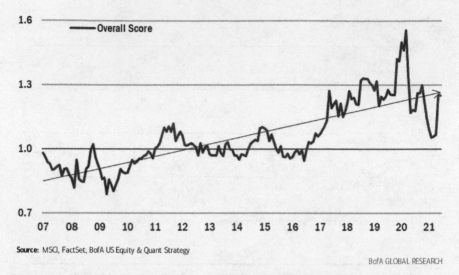

FIGURE 12.2. Bank of America calculates that "good" ESG stocks have become increasingly expensive since 2008. In early 2020, a mini-ESG bubble burst.
Courtesy of Bank of America.

of time. In other words, for a brief moment, optimism and euphoria gave way to pessimism. This valuation reset should have surprised no one. More importantly, it seems plausible that painful gyrations like this could become more commonplace. If they do, investment strategies that overweight highly ranked ESG companies will experience more volatility than their unsustainable counterparts. Heightened volatility is almost certainly not what most investors expect from sustainable investment strategies. *Heightened volatility* is a polite term for heightened risk. Sustainable investment strategies are broadly expected to generate less volatility and less risk over time, not more.

Of course, it is entirely logical for good ESG companies to trade at a premium. However, what one should want as a "total-return" investor is to own less highly rated companies on their way to becoming good, rather than good or even excellent ones that are already trading at extended premiums. "Good companies can be bad investments, and bad companies can be good investments," Ray Dalio recently reminded us. At a portfolio level, this would effectively require owning as broad an index of companies as possible—that is, John Bogle's haystack or the unscreened portfolio of Norway's oil fund. As you'll recall, it's very hard to beat large indices over time because of the way they evolve contemporaneously. Broad indices are also less subject to sectoral emotional vicissitudes.

Targeting a selection of companies that show great transformational promise seems a much more promising way to generate excess returns.

But all of this analysis is at an aggregate level. It turns out there are equally instructive findings at the individual stock level. A casual review of the most significant ESG funds quickly reveals five of the same corporate names are usually overweighted: Alphabet (Google's parent company), Amazon, Apple, Facebook (now Meta), and Microsoft (figure 12.3). While these stocks compose about 22 percent of the S&P 500 overall, they have often made up 30 percent or more of most indexed ESG and active strategies. Broadly tech and e-commerce companies, Alphabet, Amazon, Apple, Facebook, and Microsoft have relatively low carbon footprints. It's pretty easy to see how tech firms are better positioned for our unfolding energy transition than are nontech firms.

The problem with this is that each of these companies has other inherent risks—including non-negligible regulatory concerns—most of which are idiosyncratic. One such idiosyncratic risk manifested on September 10, 2021, for Apple. That's when a federal judge determined their App Store violated rules for unfair competition. Apple was ordered to allow users of their

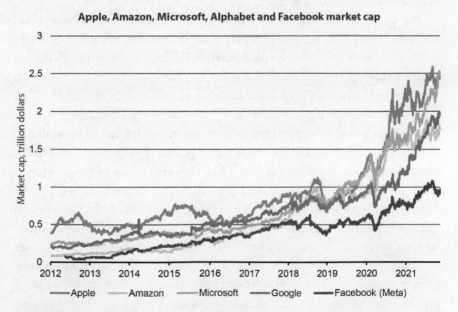

Apple, Amazon, Microsoft, Alphabet and Facebook market cap

Apple — Amazon — Microsoft — Google — Facebook (Meta)

FIGURE 12.3. Apple, Amazon, Microsoft, Alphabet (Google), and Facebook (Meta) have powered many ESG equity strategies higher. Their continued outperformance may be needed for ESG strategies to generate more alpha.
Figure by the author.

devices to include payment methods other than Apple's to buy apps. Apple shares promptly dropped by 5 percent, erasing more than $125 billion of shareholder wealth. Apple's share price had been on an eighteen-month tear-up until this ruling, nearly tripling in value since March 2020. Obviously, one of the reasons Apple's stock price had been so strong is the near tripling of ESG funds, which had overweighted their stock over the same period. A few weeks later, another idiosyncratic risk emerged, this one involving Facebook. A whistleblower claimed Facebook had knowingly harmed children in their pursuit of higher profits. As Facebook's shares plummeted, nearly $80 billion of shareholder wealth was eviscerated over one week. The threat of debilitating regulation still weighs on Facebook's future. Idiosyncratic risks increase in portfolios when stocks like Apple and Facebook are heavily overweighted. Ballooning ESG funds have had a way of increasing demand for tech stocks like these axiomatically. Discerning investors should ask what might happen to the prices of tech stocks like these if rapid flows into ESG funds were to reverse. They should also query what might happen to the performance of ESG funds if the fortunes of these five companies were to deteriorate substantially. These are just two of many idiosyncratic risks ESG investment enthusiasts must bear in mind.[4]

The Bank of America report that tracks the valuations of good and bad ESG companies also monitors corporate names that are meaningfully overweighted or underweighted versus their stated index. As it so happens, during most of 2021, a few green tech firms like Enphase Energy and some health care companies like Bio-Rad Laboratories were routinely held at eight to ten times their assigned weights in the S&P 500. This means that movements in these stocks would be eight to ten times more impactful than they would be in standard indices. At the same time, oil and gas companies like Cabot Oil & Gas and tobacco companies like Philip Morris were assigned a weight of *zero*. Once again, this is what one would expect of ESG-screened funds: deliberately overweighting and underweighting "good" and "bad" firms, respectively. But the valuations of Enphase and Philip Morris also have efficacious limits, both on the upside and the downside. The former traded as high as 175 times earnings, whereas the latter fell as low as seven times earnings. Can Enphase forever become more expensive while Philip Morris forever sinks? Yet these are the types of bets that may need to work out consistently and over prolonged periods for ESG index strategies that rely on tilts and overweights to outperform broader indices perpetually. They might outperform for a while, of course. Then again, they might just run into Herbert Stein.

Two additional ESG developments need to be watched particularly closely for their second-order effects on overly crowded trades. Both have been discussed before but need to be repeated here for emphasis. The first is the just-discussed effort underway in Europe to create taxonomies of sustainable products and the growing likelihood that similar classifications will be required in other jurisdictions. The second is an emerging suite of so-called temperature-aligned funds, such as the one mentioned earlier from Lyxor Asset Management.

The European Union's Sustainable Finance Disclosure Regulation (SFDR) was launched of necessity to combat greenwashing—the false or misleading depictions of corporations or ESG products. The SFDR is also intended to promote comparability. It requires uniform disclosures by financial market participants and advisers at an entity level and funds and separate accounts at a product level. This effort is needed, and I support it wholeheartedly. Its first designations came into effect on March 10, 2021. They are scheduled to be updated with more prescriptive classifications no later than July 2022. All in-scope products must clearly state whether and how they generically promote environmental and social goals—that is, "E" and "S" characteristics— or whether and how they target specific sustainable investment objectives, like cleaner water or carbon reduction. As we have discussed, the former are classified as Article 8 funds, the latter as Article 9 funds. Over time, how managers compose and measure the impacts of these funds will become significantly more granular. To make things harder (and only because it is Europe!), France, Germany, the Netherlands, and other EU countries are expected to impose additional local rules relating to distribution and content. How the same product is labeled in Belgium, Germany, or France may differ, in other words.

All participants agree additional rules are needed for enhanced clarity and consistency. Just like "organic" food labels, the classification of a product as sustainable or temperature aligned must mean that that product is different in some way from its generic alternatives. But many environmentally sustainable activities—like the transition to a circular economy, climate change mitigation, environmental protection, and the restoration of biodiversity—will simultaneously involve technical screening criteria and minimal, measured impacts. Unfortunately, with greater granularity and technical specificity could come temptations to game the system.

Because of this specific dynamic, temperature-aligned products in particular appear poised to heighten concentration risks. Today, a limited number of companies verifiably meet the strict definition of Paris alignment. For example, in 2021 analysts at the French asset manager Amundi reported that fewer than 5 percent of the companies in the MSCI World Index were aligned.[5] Given that the investable universe of temperature-aligned companies remains so restricted and the capital flowing into that universe so large, some unsustainably overvalued securities seem unavoidable. To express support for the goals of the Paris Agreement, a growing number of asset owners have professed to have their asset pools reflect carbon neutrality over time. These commitments are also creating mounting demand for listed products and funds that are already temperature aligned or trending in the direction of temperature alignment. This challenge of too much capital for too few qualified companies led analysts at Amundi to conclude, "At this stage, it appears difficult to build investment portfolios uniquely made up of corporates that are already aligned to a 2°C trajectory and moving towards 1.5°C."

The risks of overspecification relative to investable universes are real. This is why the chief counsel of ICI Global, Jennifer Choi, stated in her letter to the head of the International Organization of Securities Commissions the following: "We caution against imposing minimum standards that attempt to create 'categories' of products for investors rather than provide transparency for investors to make an informed investment choice. Prescriptive minimum standards attempting to define what it means to invest sustainably result in confusing disclosures for investors. Additionally, *crowding investors into product categories 'created' by regulation with a small investable universe does not further the objective of protecting investors*" (emphasis added).[6] While these words may be unwelcome, they most certainly are wise. They should also be heeded.

"Caveat emptor" is the well-known Latin expression for "buyer beware." "Caveat moderator" is the less well-known Latin expression for "regulator beware." "Emptor, caveat moderator" depicts one of my greatest concerns about ESG investing today: *Buyers, beware regulators!*

Even if a select number of regulatory jurisdictions conclude that sustainable products must contain a minimum number of temperature-aligned firms, the resulting popularity of such products could distort market valuations in ways that may not prove to be in the long-run interests of investors. The better choice would be to encourage much greater transparency and let informed investors decide which risks they wish to bear relative to their own investment goals. At no time should regulators find themselves in the business of claiming one sustainable investment product or process is better for all investors than any other, nor should they start weighing in on probabilistic scenarios. They are not prescient. No one is. As ironic as it sounds, proclamations about the sustainability of certain companies or investment strategies would more likely than not distort prices of those companies and strategies unsustainably. Attempting to promote better environmental and/or social outcomes through regulation could well become counter-productive. Official claims that certain products may well be more stable could breed a "Minsky moment"—named for Hyman Minsky, who rightly obsessed over the risks of excessive speculation—especially if their underlying assets become more volatile than expected.

Once again, John Bogle probably said it best: "Investor emotions plus fund industry promotions portends little that is lasting and good." Asset owners need to keep their emotions and investing strategies professionally segmented. While it is verifiably the case that some sectors and industries are better positioned for the energy transition, whether "sustainable products" are attractive investments at any moment in time depends first and foremost on the prevailing valuations of their underlying assets. Many outstanding ESG stocks have been trading at significant premiums. This didn't and doesn't mean they aren't good companies; it simply means they may not, at that particular time, be superior investments.

But let's not lose sight of our goals here. We aren't trying to predict the next investment bubble or instruct regulators on exactly what they should or should not do. These were never our assigned tasks. Instead, we are trying to identify promising ways finance can help promote more inclusive, sustainable growth. It is clear there are efficacious limits to what regulators can do to accelerate this journey. The law of unintended consequences is always at work. Moreover, it is most often overlooked by authorities who seek virtuous outcomes while remaining insufficiently attentive to their second- and third-order effects.

So what should investors seeking verified and verifiable double bottom lines do? We are coming to this discussion very soon—but first we must speak about a parallel universe of corporations we have not yet properly considered. This universe is not populated by publicly listed firms that are readily found in ESG ETFs or sustainable mutual funds. Rather, it consists of privately held and/or state-owned enterprises. It turns out these types of firms crucially bear upon how and whether investors can help promote more inclusive, sustainable growth. They also help us evaluate whether current ESG investments can have all the environmental and social impact many of their enthusiasts seek.

As you can surely appreciate, private firms and state-owned enterprises need to be part of the solution, not the problem. Right now, they are both.

Chapter Thirteen

LET'S SPEAK PRIVATELY

Give me a lever long enough and a fulcrum on which to place it,
and I shall move the world.

ARCHIMEDES

We have spent the bulk of our investigative efforts so far determining whether
and how the behaviors of companies could be altered and optimized through
initiatives like ESG investing and those of Just Capital. At the end of the
day, sustainable investment strategies are largely about penalizing scofflaws
and rewarding paragons, reaping financial and social rewards in the process.
We've learned that firms in the S&P 500, MSCI World, Russell 3000, or any
other index are prime candidates for direct investment influence. By overweighting or underweighting the stocks and bonds of the companies that
investors want to reward or punish, undesirable corporate behaviors may
be discouraged and beneficial social and environmental outcomes possibly
amplified. Concomitantly, better returns may also be earned. Thus, if Shell
or Lockheed Martin are engaging in some untoward activities, the threat of
removing them from certain investment portfolios may lead them to course
correct preemptively. After all, reputational and financial costs are taxing
and should ultimately persuade targeted companies to amend their ways. If
the incentives are compelling enough and markets as determinative as many
think, ESG investing should be a compelling methodology for making the
world a better place, rather like Archimedes suggests. The process of rewarding paragons should also be rewarding for investors.

But as we have looked more deeply at ESG investing and other methods,
we have also seen some of the inherent limitations of these arguments. Composite ESG scores and shifting American moral priorities do not appear to

convey complete or particularly prescient investment information. Basing indexed equity strategies on composite ESG scores alone seems unlikely to generate recurrent, long-term alpha. Moreover, investment strategies based upon composite ESG scores also appear to have minimal direct impact on corporate behavior. Unlike credit ratings, which have promoted greater corporate conformability in accounting practices, ESG scoring systems have not yet created much harmony in corporate "E," "S," and "G" practices. In particular, tilting an index through composite ESG scores does not seem to have much influence on individual corporate strategies. Corporations may want to be in an ESG index, but how much they are willing to change for inclusion remains murky. In this respect, ESG incentives and the cost-of-capital transmission mechanism upon which those incentives depend have been presumed by too many to be more potent than they actually are.

The profusion of composite ESG scores and their limited effects on corporate behavior lie in contrast to their possible direct and outsized influence on investment results, meanwhile. Incurring tracking error while failing to improve specific social and environmental outcomes is an unattractive risk–reward proposition. Furthermore, we've learned that extensive divestment efforts failed to dissuade "sin" corporations from conducting their traditional core businesses. Meanwhile, ample financial capital continued to flow to alcohol, tobacco, firearm, and gambling stocks for two compelling reasons. First, their operations remained profitable and legal; second, ubiquitous fiduciary rules encouraged it. This is also why, somewhat counterintuitively, sin shares have historically outperformed the broader market. Their relatively high dividends broadly benefited those who chose not to divest. Finally, we've learned how challenging and time-consuming the process of systematically hardwiring corporate goodness will be. TCFD and SASB are compiling valuable road maps for sustainable accounting standards. The sooner they can complete their work, the better. In the interim, though, few corporations have chosen to comply with their evolving reporting standards. This means that investors today are still some ways away from systematically identifying and discounting all material ESG risks. Admittedly, they are further along than where they were when the United Nations launched their Principles for Responsible Investment in 2005, which is laudable. But they are not where they need or want to be. Progress is being made—corporations are certainly now more "ESG aware"—but transmuting greater transparency into better investment returns and a more just, sustainable world remains a long, important work in progress.

As if these challenges were not enough, there is a further constraint on ESG's broad efficacy that we've not yet considered: the potential relevance of powerful competing economic agents, namely private markets and state-owned enterprises.

Divestment and overweighting as we've explored them so far relate meaningfully only to publicly traded companies—that is, those with shares that trade on an exchange and those whose stocks and bonds are included in widely tracked indices. But the ultimate utility of corporate engagement strategies like those of ESG, Just Capital, or B Lab for systemic environmental and social change depends upon their ability to influence the *entire* corporate ecosystem, not just a fraction of it. Thus, if promoting more inclusive, sustainable growth through corporate engagement strategies is to work, the industrial practices of all of the institutions that are its targets must be sufficiently pervasive for the global economy to be meaningfully recalibrated.

But that is not the case.

By almost every metric, listed firms compose less than half of all economic activity. Moreover, state-owned enterprises and private firms still lie well outside the ordinary reach of activist ESG influence. To date, private firms and state-owned enterprises have remained relatively immune to ESG pressures. Because listed firms represent such a small segment of global commercial activities, ESG's current lever is not long enough, nor is its fulcrum sufficiently secure to protect the earth's climate or alter social outcomes consistently and systemically on its own.

All of this is why we must speak privately.

While the primary targets of corporate engagement—the fifty thousand or so corporate names that are included in the most widely followed indices—are the most visible components of the global economy, they are far less influential than is commonly understood. Consider the U.S. economy. Of the thirty million businesses incorporated in the United States, *only 0.0125 percent are listed* on an exchange or access the public bond market. For universal impact, activists and ESG enthusiasts need to develop additional strategies for engaging the remaining 99.9875 percent. Part of this challenge is a function of firm size, of course, but only part. Globally, corporations with fewer than one hundred employees outnumber those with more by a margin of seventy to one. While thirty-five million Americans work for Russell 3000 companies, this still amounts to less than one-quarter of the 155 million Americans who go to work every day. If promoting fair wages, greater diversity, and employee satisfaction are top goals—to list just a few desirable objectives—proponents for improving labor market outcomes must also find a way to reach employers of the other 77 percent.

Authoritative, comprehensive data on private firms is hard to find because corporations that do not register to sell securities need not file detailed activity documents with organizations like the Securities and Exchange Commission. This means that more than 99 percent of all U.S. corporations never report their cash flow, inventory turnover, investment, or employment data, let alone granularity on their scope 1, 2, and 3 emissions. In one of the more recent studies comparing public firms with their private counterparts, researchers at Harvard and NYU estimated that privately held firms were responsible for 70 percent of all nongovernment employment, 60 percent of all sales, and 55 percent of all nonresidential fixed investment. They also estimated that private firms account for more than 50 percent of all pretax corporate profits.[1]

Upon further reflection, though, this data should not surprise anyone. Many of the largest privately held companies in the world are household names. These include accounting and advisory giants like Deloitte and Ernst & Young, with 330,000 and 290,000 employees, respectively; luxury brands like Rolex and Lacoste; the financial giants Bloomberg and Fidelity Investments, with 60,000 employees between them; Dell, with nearly $60 billion in revenues and 110,000 employees; and the confectionery giant Mars, the supplier of my beloved Twix and Snickers bars and the employer of more than 130,000. Some industries are dominated by private giants. Of *Forbes*'s two hundred largest private companies in America, sixty are in the food business, including fourteen of the top twenty-three.[2] Two other well-known names—Cargill and Koch Industries—are among the world's largest commodity and energy engineering companies. Were we to assign Cargill and Koch conservative price–earnings multiples, given their annual revenues exceed $100 billion each, both would rank among the top twenty-five most valuable companies in the S&P 500. The Dutch energy and commodity trader Vitol is even larger than Cargill and Koch, meanwhile, with reported annual revenues exceeding $250 billion. Vitol's global sales are higher than Google's and Samsung's. Hundreds of private companies have thousands of employees; dozens have hundreds of thousands. When you add all of this together, you realize a largely underappreciated fact of commercial life: the activities of private firms matter much more in the global corporate ecosystem than do those of their public counterparts.

Most private companies are not household names. Often, this is intentional. Many significant private companies eschew public attention. Thus,

Hilcorp may be the largest producer of oil and gas you've never heard of. Unless you are in the oil and gas business, there is no reason that you would have. Yet Hilcorp ranks near the top of all oil producers in Louisiana, Ohio, Pennsylvania, Texas, and Wyoming. They are also the largest oil or gas producer in Alaska and the San Juan Basin. Hilcorp earned top billing in Alaska after buying all the Alaskan upstream and midstream assets of BP in 2019 for $5.6 billion. In this landmark transaction, BP divested all their interests in the giant Prudhoe Bay field, as well as their 48.4 percent stake in the Trans-Alaska Pipeline System. In the process, BP shed a considerable amount of financial and reputational risk. As you would expect, BP won plaudits from the investment community for their decision; their stock rose more than 8 percent in the first few weeks that followed this announcement. But BP's sale certainly did not mean that Alaska's oil would somehow stay in the ground. Meanwhile, Hilcorp plans to make the most of its multi-billion-dollar purchase, as you might expect. And while ESG investment enthusiasts cheer BP, they have no meaningful way to express their displeasure with or influence over Hilcorp, the other side of the transaction. Hilcorp has also so far profited from their purchase.

I mentioned Shell and Lockheed Martin at the beginning of this chapter for a reason. Both recently used the same playbook as BP, shedding controversial assets that seemed to weigh on their enterprise values. They sold them to privately held companies or to other firms that could not be branded negatively. Specifically, Shell and Lockheed spun off carbon-intensive energy interests to show critics they were making some progress toward the goals of Paris. Shell sold their significant oil sands holdings in Canada to a subsidiary of Canadian Natural Resources for $7.25 billion in 2017 and a portfolio of onshore, upstream assets in Egypt for $646 million in 2021. Lockheed sold their energy services business to a privately owned firm, TRC, in October 2019. In each of these cases, shareholders cheered the results, while the corporations that took over the controversial assets and businesses incurred no corresponding penalties. The global economy also became no greener or sustainable on a net basis, even though investors responded on balance as if it somehow had. These actions help illustrate why Reclaim Finance wants to punish all businesses that sell gross carbon assets instead of closing them down and writing them off. Moving ownership of Canadian oil sands from one pair of hands to another generates no net benefit for global carbon accounting.

Salacious stories impugning the Koch family and their vast influence over large swathes of the U.S. economy and public policy are standard media affairs. Their right-leaning—perhaps more accurately, libertarian—viewpoints have made them an irresistible target for left-leaning publications. *Rolling Stone* magazine largely set the tone in a 2014 expose: "The Koch family's lucrative blend of pollution, speculation, law-bending, and self-righteousness stretches back to the early twentieth century. . . . Charles and David Koch control one of the world's most enormous fortunes, which they are using to buy up our political system; what they don't want you to know is how they made all their money."[3]

But how the Koch brothers make their money is quite simple and straightforward. Koch Industries flourishes by providing a vast array of products and services that consumers need and consistently use. Fuel for trucks, cars, and jets; tempered glass for construction and transport; fertilizer for more abundant crops; surgical and medical devices, including insulin infusion pumps; paper products ranging from napkins and disposable towels to cups and toilet paper; fabrics for combat uniforms and first responders. Like other so-called sin companies, Koch is well positioned to provide products and services people are certain to consume whether others approve of their consumption or not. Note too, Koch is particularly well positioned to provide products that have been historically supplied by now-sanctioned Russian firms, like oil and fertilizer. Is it not better to buy these products from Koch Industries than Putin? Meanwhile, what you will not have read in *Rolling Stone* is that Koch Industries and their affiliates have simultaneously won 1,200 industry awards since 2009 for environmental excellence, community stewardship, safety, and innovation in sustainability. In fact, in April 2021, Koch Industries won the Environmental Protection Agency's top Energy Star Partner of the Year Award for protecting the environment through improved energy efficiency. Since 2015, Koch has also invested more than $1.5 billion to promote energy conservation. Today Koch recycles and treats 91 percent of their waste products, more than many companies who claim membership in the circular economy. Koch Industries is also meticulous about following the law. Add this all up, and you can be sure Koch Industries is not going out of business anytime soon. As they are crucial to maintaining pressure on Putin, I suggest we thank them rather than persecute them. You can also bet that, should the shareholders of Exxon, Chevron, or any other public energy producer be forced to exit their oil and gas operations, companies like Hilcorp and Koch will be ready to step in at the right price and bring those carbon assets to market. As long as it is legal, financially attractive, and strategically important, Hilcorp, the Koch brothers, and other private firms like them are primed to supply the oil and gas that consumers continue to use. None of this is to heap judgment on Hilcorp or Koch, positive or negative—though it must be mentioned Hilcorp has been named to *Fortune* magazine's "100 Best Companies to Work For" for seven straight years. Instead, my primary point is that every molecule of oil held by public corporations could be liquidated—something many activists covet—yet all those carbon assets could be snapped up by private purveyors and still end up being consumed. Legally.

There are obvious operational limitations to optimizing the trajectory of the global economy through publicly listed companies alone. Privately held companies play outsized roles in a wide range of industrial, environmental, and employment practices, including the production of carbon-based fuels. Since 2010, the private equity industry has invested $1.1 trillion in the energy sector, double the combined market value of Exxon, Chevron, and Shell. Private energy companies are also ready to do much more. "We have an appetite to acquire," Brian Gilvary, the head of Ineos Energy, the arm of a private UK chemicals company and a former executive of BP, recently said. Ineos bought Hess Corporation's oil and gas assets in Denmark for $150 million in the summer of 2021. Gilvary went on, "We're a private company with private shareholders, but we still have to operate in a way that is in line with what governments, banks and investors want to achieve."[4]

Beyond publicly listed and privately run firms, there is yet another segment of the global economy that will limit the impact of pressuring domestic energy companies to change their ways: state-owned enterprises. Shell and BP; Exxon and Chevron; and Hilcorp, Ineos, and Koch Industries are all nongovernmental organizations. Each occupies an important component of the global energy complex. But China's National Petroleum Corporation, Saudi Arabia's Aramco, Russia's Rosneft, Brazil's Petrobras, Venezuela's PDVSA, and Iran's National Iranian Oil Company are all government owned. When it comes to the future of oil and gas production, taken together, these state-owned enterprises are far more determinative.

More than three-quarters of the world's known oil reserves are not in private hands; they are under government control. This means that state-owned oil companies are three times more important than their privately run counterparts. With annual revenues of $450 billion, Sinopec is the world's largest oil company and nearly twice the size of Exxon, the largest U.S. producer.

Sinopec's sister, CNPC—China's second-largest state-owned oil company—
is about the size of Exxon and Chevron combined. Obviously, it will not
be possible to alter the globe's production and distribution of oil and gas
meaningfully—making it cleaner and greener—without convincing the
governments of Brazil, China, Iran, Russia, Saudi Arabia, the United Arab
Emirates, and Venezuela that they, too, must be part of the solution. Just like
European gas imports, U.S. importation of Russian oil reached a record in
2021, surpassing imports from Saudi Arabia. As U.S. producers pulled back,
America's most pernicious adversary and the world's dirtiest oil producer—
Russia—had to step in to sate U.S. consumer needs. Shortsightedness in
Europe and the United States meant revenues for Russia's state-owned
energy champions and war funds for Putin's decimation of Ukrainian cities
and citizens.

In 2017, Sinopec expanded into Africa by purchasing 75 percent of Chevron's South African assets. Just as Hilcorp's purchase of BP's Alaskan assets
benefited BP's ESG rating without making the world any greener, Sinopec's
purchase allowed Chevron to score highly with energy activists. At the same
time, the state-owned purchaser paid no reputational price. "We spend all
this time focusing on BP and Shell," Angela Wilkinson, the head of the
World Energy Council, caustically observed. "What about Saudi Aramco
and ADNOC [Abu Dhabi National Oil Company]? We see pressure on a
small subset of listed oil companies. But it's not at all a realistic picture of the
overall energy system."[5]

The leading energy consultancy Wood Mackenzie estimates that BP, Chevron, Exxon, Shell, Total of France, and Eni of Italy—all listed companies—
have disposed of more than $28 billion in energy assets since 2018. Most
of these dispositions went to private or state-owned hands. Wood Mackenzie projects further disposals of more than $30 billion in energy assets
in the years to come. The total value of oil and gas assets that may be up
for sale from listed companies over the next decade could be more than
$140 billion. The quickest way to shrink one's carbon footprint and claim
temperature alignment is to shed physical assets. But such sales do nothing to reduce emissions; they merely move carbon assets from scrutinized
balance sheets to less scrutinized ones. A 2021 ruling from a Dutch court
insisted Shell cut their absolute carbon dioxide emissions far more quickly
than planned. Shell's subsequent decision to delist in the Netherlands was a
predictable response. It's likely that most of Shell's disposed energy assets will
still end up in state-owned or privately held companies, but they can now

proceed at a more measured pace. Of course, if they do, Shell may end up greener in the end, but the world almost certainly will not.

A recent OECD report zeroed in on the double challenge state-owned enterprises present for optimal environmental, social, and governance outcomes. When it comes to "G," state-owned enterprises appear uniquely problematic. It turns out that governments are not very good at governing themselves. Nearly half of all state-owned enterprises report losing an average of 3 percent of annual profits to corruption or other irregular practices, a multiple of private sector infractions.[6] Though state-owned enterprises compose about one-fifth of all global economic activity, they account for more than half of all known corporate corruption. The OECD estimates that 1.5 percent of state-owned enterprise expenses are dedicated to detecting or preventing corruption. Improved governance standards are needed globally—but they would make the most significant impact if they were applied by governments to their wholly owned utilities, natural resource, finance, transportation, postal, agricultural, and communications companies.

Publicly listed companies receive the most significant amount of scrutiny from activists because they are the most visible. Privately held firms are no less law-abiding than their listed counterparts. Still, the playing field is unlevel because they operate further away from public glare and are subjected to much less reporting rigor and ESG capital allocation scrutiny. This inconsistency allows private firms to exploit market inefficiencies that divestiture pressures and trends invariably create. State-owned enterprises are in a league all their own from both operational and legal standpoints. They are held to account only when their governmental owners decide it is in their best interest to do so. Moreover, given that state-owned enterprises are unusually inclined toward bribery and other forms of corruption, their peculiarities pose a unique challenge to global sustainability accounting standards. Putin's 2022 oil revenues will likely exceed those of 2021, moreover. This too is infuriating.

Why have I gone to such lengths discussing private firms and state-owned enterprises? To make this point: *If net zero is a global aspiration, we must go much further than encouraging BP, Exxon, Shell, Total, and their other publicly listed counterparts to clean up or abandon their core energy businesses.* They are a very, very small part of the much larger energy corporate ecosystem. If we want businesses to accelerate human development, reduce their emissions, and improve working conditions throughout the world, we will need to focus on privately held companies and state-owned enterprises at least as much. Even more importantly, we must address the issue of energy demand.

The threats of divestment and the promise of higher valuations for publicly listed companies taken by themselves implicate a surprisingly small segment of the global economy, especially in the energy sector. While this means strategies for reducing carbon emissions must become more comprehensive, it is also true for other social priorities as well, including workforce diversity and economic mobility. Now that we more fully appreciate the role private and state-owned oil companies play, it is also increasingly apparent that the globe has little prospect for achieving the goals of Paris without significant changes in policies that apply to *all* economic agents. One principal lesson of the past year is that reducing oil supplies while demand remains unabated or continues to rise undermines self-sufficiency and impacts the poor the hardest by sending oil and gas prices higher. Had we been more thoughtful about ongoing demand dynamics, painful energy price hikes and our senseless reliance on Russian oil and gas could have been avoided.

Does all this mean we should stop looking to publicly listed corporations to minimize their environmental, social, and governance risks? No, it does not. It merely raises concerns about how effective publicly listed–only activist campaigns are. It also raises questions about how much more we must do to promote inclusive, sustainable growth comprehensively.

But this is not the only challenge of campaigns that favor divestment as a tactic for forging progress. Divestment arguments have always had another weakness. Not only does the threat of share divestment not work as hoped; the act of divestment is also inherently counterproductive. When it comes to changing corporate behaviors, it is more effective to be an inside shareholder fighting for change than an outsider carrying placards and shouting slogans on the picket lines.

Much more effective.

Chapter Fourteen

FIGHT OR FLEE?

We have become . . . the stewards of life's continuity on Earth.
We did not ask for this role, but we cannot abjure it.
STEPHEN JAY GOULD

We believe investors can only effect lasting change when
they work as active owners.
ENGINE NO. 1

The decision to own or not own the shares or bonds of a publicly listed enterprise is consequential across three dimensions. Two were discussed in an earlier chapter: ethical alignment, or values, and investment returns, or valuations.

Asset owners have the right and perhaps the moral obligation to maintain stakes in companies that reflect their core principles and beliefs. Those who choose to do so enjoy a peace of mind that ethical consistency alone confers. However, exercising this right can and often has come at the expense of risk-adjusted returns, that is, the valuations dimension. No principled, socially responsible investor should expect that their decision to exclude certain companies or industries from their investment universe for normative reasons will axiomatically improve their investment results. Ethical screenings may or may not improve returns. Neither outcome is assured. If anything, recent evidence suggests forgone returns may be more likely. Sin stocks have outperformed their counterparts over an extended period, and the share prices of oil, gas, and other extractive companies significantly outperformed their greener counterparts in 2021. Many tech stocks that drove ESG outperformance prior to 2022 have also massively underperformed of late. Past performance is never a guarantee of the future, of course. One thing we do know is that the more concentrated a portfolio becomes, the less diversification benefit it contains. Holding a large number of stocks and selectively divesting presumed under-performers introduces tracking error. If one's divestiture strategy is highly sophisticated and successful, one may be able to avoid losses and generate significant excess returns. Axiomatically, the converse is also true.

But divestiture also entails a third, too-often underappreciated dimension: *the retention or elimination of one's right as a shareholder to participate directly in corporate governance.* Shareholders are the ultimate owners of a company, after all. Class A or common stock shareholders have meaningful governance oversight powers. They can put forward resolutions, participate in the election and removal of board members, and contribute to the strategic direction companies may take. They also have the right to receive any dividends corporations pay out, as the Dodge brothers memorably demonstrated. Boards run public companies, but shareholders ultimately decide who sits on corporate boards. Whether they realize it or not, asset owners help drive the bus.

For ESG-oriented or any other investors looking to alter corporate behaviors while simultaneously improving their returns, divestiture intuitively seems self-defeating. From the standpoints of both social utility and investment return, it would seem vastly preferable to stay engaged in corporate oversight, fighting for one's interests, beliefs, and goals than to walk away. Working to help secure improved ESG metrics or better temperature alignment, ceteris paribus, should also result in higher equity valuations. On its face, divestiture precludes such value-added engagement. Divestiture rules out the opportunity for owners to reform companies from within, as well as any rewards that may come from positive transformation.

Helpfully, recent academic research on the relative value of shareholder engagement versus divestiture confirms what intuition suggests. Eleonora Broccardo of the University of Trento, Oliver Hart of Harvard, and Luigi Zingales of the University of Chicago found that neither divestiture nor consumer boycotts have the hopeful impacts their proponents seek. Instead, their research found that remaining an engaged, active shareholder is a superior method for promoting desired objectives: "Exit is less effective than a voice in pushing firms to act in a socially responsible manner. Further, individual incentives to join an exit strategy are not necessarily aligned with social incentives, while they are when investors are allowed to express their voice."[1] Stated more succinctly, it is better to fight than to flee. Their research on divestiture and boycotts versus engagement focused primarily on climate concerns. They believe their analysis applies to all social and environmental issues, however: "Individual shareholders have the incentive to vote on issues in a socially optimal way, and their engagement can lead to more efficient outcomes."

These conclusions are highly significant for the broader ESG debate. Not only do they suggest divestiture is a less effective strategy than engaged share ownership, but they also propose that to exercise one's convictions, it may be better to actively *overweight* the companies most in need of redirection so one can have a more prominent voice. As we've learned, many ESG strategies advocate the opposite stance; by screening out unwanted companies, some asset owners effectively flee rather than fight. Broccardo, Hart, and Zingales's findings also raise an important, related question: if exit is relatively ineffective, why do so many activists and informed investors choose exit over engagement? One reason may be that many of those activists aren't shareholders to begin with. Not able to vote themselves, they might presume it's best to persuade others to join them on the sidelines, as conscientious objectors. This makes no sense, however. Not only does divestment mute potentially impactful participation in shareholder decisions, it also leaves those aligned with existing managements more firmly in control. Another equally plausible explanation for preferring divestiture may be that its proponents don't believe their votes matter. For them, the entire system may appear "rigged," and they want no part of it. However, with the recent evidence uncovered by Broccardo, Hart, and Zingales, such a position would need to be reconsidered. Exit is not only less effective than voice; it actively thwarts the possibility of beneficial corporate transformation. Effective activists should be vocal shareholders and active insiders, not outsiders shouting barbed criticisms from the sidelines. Finally, divestiture's proponents may not trust their proxied adviser to vote correctly, or they may not have been offered convenient options by their custodian or investment advisers for direct participation. Significant effort and attention are needed to conduct investment stewardship effectively. To lack philosophical alignment with one's proxied choice would be even more detrimental than not to vote. One's proxied voice may contravene one's actual preference, advertently or inadvertently.

You'll recall from our earlier discussions that the Norwegian oil fund retains a significant professional stewardship team. They use it proactively, engaging companies on issues like water management and children's rights to other strategic and tactical issues about which they are most deeply concerned. The Norwegian oil fund has explicitly chosen to remain an activist shareholder, with divestment only as a last resort. They have done so because they have experienced firsthand how such efforts make a positive difference. They want to be in the room where it happens. Most importantly, they want the corporations they own to maximize shareholder value over time; after all, if you must own a company in perpetuity, you want it to be the best it can be.

Other investors may be too small or inexperienced to be similarly engaged, though. They may not see themselves as long-term owners. Intentionally or unintentionally, most institutional and individual shareholders second their voting rights to their asset managers, effectively empowering them to be their advocates for better returns as well as other interests.

For investment managers, this secondment confers a tremendous responsibility. It also means stewardship has become yet another tool in their toolbox for attracting and retaining customers. How so? If investors feel morally and strategically aligned with an asset manager's voting record, they should also feel more inclined to give that manager their business. But, of course, the converse is also true. If an investor feels their investment product provider is not representing their views and interests properly through shareholder resolutions and proxies, they could and obviously should move their money elsewhere. Asset *owners* effectively empower their asset *managers* at the end of the day. But it's the voice of asset owners that most needs to be heard, especially if that voice helps maximize long-term value.

All this helps explain newfound interest in and appreciation for the power of shareholder voting that has been demonstrated by Engine No. 1, the investment firm that successfully mounted a challenge to Exxon's board. In a spirited play for similarly aligned asset owners, Engine No. 1 has promised to use the tools of capitalism to improve capitalism more aggressively. In a pitch for like-minded asset owners, they unveiled a new investment philosophy in September 2021. In it, they criticized the current approach of most ESG funds and proposed an alternative strategy, which they call the "Total Value Framework":

The financial returns by most ESG strategies have been equivocal at best. . . . Far from changing corporate behavior and creating a better world, some ESG funds have served the narrower and more self-indulgent purpose of making investors feel better by excluding obviously "bad" companies. . . . This failure to achieve goals commensurate with the ambition is partly the legacy of an approach still focused on moral purity rather than impact. Absent active managers who can produce superior returns, investors have focused their attention on funds that mirror indices modified by the exclusion of stocks with unfavorable ESG ratings. . . . What's needed now is a radical new research-based approach that integrates nontraditional but financially material ESG data, methods, and systems into the traditional analysis. Without such a change, ESG investing is unlikely to harness the power of capital needed to address systemic challenges like climate risk and human rights. [Moreover,] without more convincing evidence of a link between ESG and performance, screening or exclusionary strategies will continue to dominate. . . . As a result, we face the danger that the passion, energy, and vision that have gone into the ESG movement will soon dissipate into a cloud of uncertainty and confusion.[2]

In short, Engine No. 1 says divestment isn't enough to bring the changes modern capitalism most needs—but they do not stop there. They further claim highly popular ESG strategies won't be effective either. After all, they claim, such strategies have already excluded the companies that most need to change. Instead, they argue, what is needed are visionary, activist investors like themselves who know how to engage managements in ways that fundamentally transform companies and the economic system to create lasting value and resilience. Undoubtedly, their recent experience with Exxon is intended to serve as the poster child for their ambitious approach. They fought for change from within, and they won. As in Exxon's case, however, they will need similarly motivated shareholders to fight alongside them to succeed. To this end, they recently launched an ETF to garner more shareholder support and put their new Total Value Framework into practice. Named the "Engine No. 1 Transform 500 ETF" and listed under the symbol "VOTE," its stated investment objective is to track the Morningstar U.S. Large Cap Select Index before fees and expenses. While this sounds much like many of the ESG strategies they deride, beneath their unambitious return target is a transformational promise: join us and become part of a large-scale campaign to reform the corporate world one company at a time. Together, Engine No. 1 contends, activist shareholders can and will forge genuine sustainability and greater long-term prosperity.

VOTE had a market capitalization of about $290 million at the end of 2021. Obviously, with such a small volume of assets, Engine No. 1 must continue to rely upon their considerable powers of persuasion to pass any shareholder resolutions. Their approach won't be able to succeed without the support of many more like-minded investors. Among potential supporters are large index product providers, including Vanguard, BlackRock, and State Street Global Advisors. As we have already learned, these institutions neither work together in any explicit manner nor operate as a "bloc." This said, added together, they control about 18 percent of all publicly listed shares in the United States. Engine No. 1's prior success in transforming Exxon's board was made possible only by the support they received from larger voting blocs. Getting large managers behind their nominees proved pivotal. It also highlighted the growing importance of large index asset managers in corporate governance. While some welcome this growing power as a harbinger of better corporate decision-making, others fear it, insisting it must be kept in check. The influence of large index asset managers on corporate decision-making is undoubtedly increasing. Given competitive trends—including the ongoing power of innovation as well as pressures for lower pricing, which is possible only through scale—they are also likely to grow further.

Professor John C. Coates of Harvard Law School belongs to the camp that fears the growing influence of the larger asset management firms. In an article published in conjunction with Harvard Law School's Program on Corporate Governance, Coates highlighted three mega-trends reshaping corporate decision-making: *globalization, private equity,* and the *rapid growth of index investing.* In his considered judgment, all three trends threaten to entangle business with the state more deeply and permanently, solidifying powers within a small group of individuals. In fact, in his estimate, there will ultimately be no more than twelve "voices" that matter.[3] If these twelve routinely choose to vote similarly, Coates worries, they could ultimately sway every vital decision taken by boards, proxies, and resolutions.

Coates's "Problem of Twelve" title is self-evidently misleading. For one, he suggests the votes of organizations like Vanguard, BlackRock, and State Street are controlled by "three people" when, in fact, the input of hundreds of qualified, properly aligned, and highly engaged professionals contribute to stewardship decisions at each of these firms. In fact, no one person at Vanguard, BlackRock, or State Street controls any corporate vote. But he is not wrong to observe that corporate voting decisions are undergoing important new dynamics. The significance of voice over exit and how varying stewardship philosophies can impact corporate outcomes increasingly deserves to be part of the investment decision-making process that every asset owner makes. For this and other reasons, it is worthwhile spending a bit of time on how stewardship practices are evolving. For example, as prominent index providers have grown, each has added additional stewardship resources and amended their voting principles and rules of engagement with corporations.

Working separately, the stewardship teams at Vanguard, BlackRock, and State Street directly engaged thousands of companies on substantive issues relating to their clients' interests in 2021. These engagements took the form of in-person and virtual meetings as well as targeted letter writing. Their corporate engagements spanned more than seventy countries. Over the same period, Vanguard, BlackRock, and State Street collectively voted more than

450,000 times on proposals involving at least ten thousand companies. This amounts to an average of 1,600 informed votes every working day of the year. Together, these actions represented significant increases over 2020. Steward-ship engagements between 2019 and 2021 at Vanguard, BlackRock, and State Street grew at the astonishing annual rate of 50 percent. If this has been the work of just three people, they must be three very remarkable people.

In addition to these engagements and votes, all three firms have made concerted efforts to refine and publicize their evolving stewardship pri-orities. It is easy to understand why they would. Greater transparency on governance philosophies can help companies anticipate how their pend-ing resolutions and director elections are likely to be adjudicated. The goal of most professional stewardship efforts is to keep the corporations they engage on optimal, *evolutionary* growth paths. Large, professional steward-ship teams are not trying to play "gotcha" with corporate boards, nor do they have anything to gain by stealth attacks. Given their goals are to maximize long-term shareholder value and interests *in perpetuity*, they logically prefer to engage corporations constructively, consistently, and transparently, over long periods. This said, there are still substantive differences between their stewardship approaches.

State Street's stewardship mission is "to invest responsibly to enable pros-perity and social progress." To do so, State Street uses a proprietary ESG scoring methodology known as the "R-Factor." They have designed their methodology in accordance with PRI and SASB principles. Historically, State Street has targeted corporations with poor R-Factor scores because they believe that it is those firms that must change the most to perform better over the longer term. If progress is not made across the board—across all indus-tries and corporate types, but most especially with laggards—prosperity and social progress are less likely. As they note on their website,

> As near-perpetual holders of the constituents of the world's primary indices, we take a value-based approach and use our voice and vote to influence com-panies on long-term governance and sustainability issues. Our approach to stewardship focuses on making an impact. Accordingly, our stewardship pro-gram proactively identifies companies for engagement and voting to mitigate ESG risks in our portfolios.[4]

While weak R-Factors are one sign of probable State Street engagement, the State Street stewardship team has been particularly front-footed on the issue of gender diversity on boards. State Street has done this through their highly publicized "Fearless Girl" campaign. As a result, many of State Street's most pointed engagements with companies in recent years have been with regard to the "S" and "G" categories, with the goal of promoting greater gender diversity.

Like State Street, BlackRock's stewardship team also publishes an annual statement of their principles and corporate voting guidelines as well as their annual voting record. Their stated objective for 2021 seems somewhat aligned with State Street's, at least in spirit: "the pursuit of sound governance and sustainable business practices that promote long-term value creation." That said, their published 2021 summary of expectations suggests they might apply their appointed voice somewhat differently in certain areas. BlackRock's investment stewardship team publicly listed three priorities as paramount: (1) board and workforce diversity consistent with *local market best practice* (emphasis added); (2) a demonstrated understanding of key stakeholders' interests; and (3) *specific plans to align business activities with the global goal of net zero greenhouse gas emissions by 2050* (again, emphasis added).[5] BlackRock's explicit embrace of stakeholderism, as well as the goals of Paris as stewardship principles, not only distinguishes them from State Street in style and substance but also sets them apart from the stewardship priorities of Vanguard.

While also stating their stewardship practices are dedicated to the promotion of the long-term success of companies and their investors, Vanguard has published four governance principles that seem relatively generic by comparison: (1) board composition; (2) oversight of strategy and risk; (3) executive compensation; and (4) governance structure. Climate concerns for Vanguard may fall under (2) but are not linked in any obvious way to the Paris 2050 net-zero commitment. Based upon their public and printed records, Vanguard's primary stewardship focus instead appears to be governance.

While these principles and goals may appear socially and environmentally ambitious to some, and even though they differ in scope and priorities among themselves, as a group over the years, Vanguard, BlackRock, and State Street have received more pointed criticism for voting *in support* of management than against.[6] Data from 2000 to 2018 reveals that their collective voting records were aligned with management preferences more than 90 percent of the time. Their historic support for shareholder-backed proposals has also been somewhat below what other investment firms regularly registered.[7] More recent voting records suggest this historic alignment with the status quo may be changing, but only slightly. BlackRock voted against or abstained from supporting director-related proposals or in favor of shareholder proposals involving 6,451 unique companies in the first six months of 2021. In 42 percent of shareholder meetings over the same period, BlackRock's stewardship team also voted in some manner against management's wishes. BlackRock's verbal commitment to engage companies more forcefully on climate, diversity, and other matters deemed essential to their sustainability goals appears to be translating from word to deed, but this change is far more evolutionary than revolutionary. As such, concerns that the biggest asset managers are aggressively practicing "woke capitalism" or reigning over many corporate boardrooms seem overblown.

Votes are not the only, or perhaps even the most effective, way modern stewardship practices bring about corporate change, however. Behind the scenes, unrecorded engagements often prove more impactful. Votes imply black-and-white positions, when, in fact, positions on multiple company issues more often than not involve varying shades of gray. Most corporate positions need to be carefully tracked over time for true progress to be seen. Though progress may appear slow, many corporate governance issues, like board composition, take years to enact. Other forms of investment stewardship extend beyond quiet diplomacy. For example, in the spring of 2020, as COVID-19 quickly snarled global maritime activity, the stewardship team at Fidelity International mobilized to defend the interests of hundreds of thousands of seafarers who had suddenly become stranded aboard their vessels, separated from loved ones for months on end. After assembling a global consortium of investors and gaining the attention of the United Nations, Fidelity and their sympathizers prevailed upon more than sixty countries to designate seafarers as key workers. Within fifteen months, two hundred thousand maritime professionals found their troubled lives had broadly normalized. Stewardship officers at Fidelity International used their considerable financial powers of persuasion to change hundreds of thousands of lives for the better. No votes or shareholder resolutions were involved.

Similarly, when South Korea's largest electric utility firm, KEPCO, announced plans to continue to invest in new coal-powered electrical plants across Southeast Asia and South Africa, members of BlackRock's Asia-based stewardship team became more engaged. They proactively reached out to KEPCO's management team with calls and letters. They kept up their engagement until October 2020, when KEPCO formally announced an anti-coal policy. In that announcement, KEPCO vowed not only to end all future coal-fired projects, but they also revealed all their existing coal-fired power interests would be terminated by 2050, in line with Paris. The Fidelity maritime worker and BlackRock KEPCO victories demonstrate the important yet largely underappreciated impact of large, global, professionally engaged stewardship teams beyond mere voting. It is also unlikely that value-added measures like these could be successfully mounted by Engine No. 1 or some other small yet sustainably minded investment firm.

While we have clearly established that it is better to fight than flee, we've also seen how the growing influence of large index providers generates anxieties. "I'm concerned about the amount of power they wield, even if they wield that power well," the Morningstar researcher John Rekenthaler wrote recently of Vanguard, BlackRock, and State Street, referencing their accelerating inflows of ESG funds.[8] And a bit earlier, Rekenthaler's concerns were voiced by another, more surprising insider: none other than John C. Bogle himself, the father of index investing.

In a 2018 *Wall Street* op-ed, Bogle wrote, "If historical trends continue, a handful of giant institutional investors will one day hold voting control of virtually every large U.S. corporation. I do not believe such concentration would serve the national interest."[9] However, Bogle went on to say there were no obvious solutions to this concern. He quoted Professor Coates's praise for the democratization of finance, which has been largely wrought by indexing and cheaper investment solutions: "Indexing has created real and large social benefits in the form of lower expenses and greater long-term returns for millions of individuals investing directly or indirectly for retirement. A ban on indexing would not be a good idea." In other words, Bogle believed as strongly about the social value of index investing in his last days as he did throughout his life. Society benefits tremendously from the provision of low-cost investment products; profits stay in the pockets of asset *owners* rather than their managers. At the same time, he worried about voting control becoming too concentrated.

For all these reasons, it should perhaps be welcome news for all that one of the largest index providers—BlackRock—announced a new policy in October 2021 to help ensure that the users of their index products employ the voice they themselves most want to express. As of January 2022, owners of a number of BlackRock's indexed products have been able to select from a range of third-party proxy voting policies. If asset owners don't like the vote choices offered, they may also choose to vote directly on select resolutions or in specific corporate board elections themselves. These third-party proxy choices include "status quo" options but also offer other default options such as Catholic screens and more aggressive sustainability frameworks. Such asset owners can also choose, should they wish, to continue to rely upon BlackRock's investment stewardship to speak for them. The choice is theirs. BlackRock's voter choice option—the first by a large indexer—should help ease concerns expressed by critics on both the left and the right that Black-Rock had somehow usurped their clients' right to speak for themselves.[10] In short, BlackRock is taking necessary steps to help their clients express their own views and preferences. As you might expect, this decision was met with acclaim. "BlackRock's voting choice initiative supports Washington State Investment Board in achieving our financial and corporate governance goals," Chris Phillips, the WSIB's director of public affairs, affirmed at the time of the announcement. "This technological and operational advancement is helping us create alignment between asset stewardship priorities and investment outcomes." Given the likely popularity of choice, it is hard to imagine other managers large and small will not soon follow with stewardship options of their own. With choice and voice combined, corporate boards should also find their shareholders more involved with many of the crucial decisions they must make. Ideally, corporations will respond with heightened attention to asset owner education, better long-term decision-making, and more consistent value creation.

We have come a long way in this chapter, identifying several appropriate actions that investors who seek verifiable, double-bottom-line outcomes from their investments should take. Importantly, we've learned that they should stay invested rather than divest. We've also learned that they should choose investment products managed by firms with stewardship teams that either align with their values or offer them sufficient choice to exercise their voice and promote the outcomes they most desire. Equally important, we've learned that investors should consider allying with other sophisticated, well-informed, active managers who help identify and successfully engage underperforming companies in need of transformation. Great societal and financial value can be created by any proven process that turns ugly corporate ducklings into swans. Overweighting such firms when so much industrial and operational change needs to occur should also generate considerable financial reward. Further, engaged shareholder oversight may also be humanity's best hope for promoting the type of inclusive, sustainable growth we first dreamed of in the opening chapters. Professional stewardship by informed, active shareholders focused on lagging performers seems a highly promising investment strategy for delivering double-bottom-line results consistently. Engaged stewardship is also almost certainly the best way for strategies like the Shared Value Initiative and Alex Edmans's "pie economics" to propagate at scale.

This discussion has also allowed us to make an important distinction: that between the stewardship approaches of large asset management firms that exercise fiduciary duties on behalf of their clients, versus those managers—often smaller in size—who exercise more aggressive shareholder activism. The latter more often than not seek specific environmental and social outcomes as ends in themselves, thus de-emphasizing broader shareholder concerns. No one should presume that doing the former always aligns with the latter; it doesn't. Complex trade-offs between various stakeholder goals—relating to labor, social, or environmental goals—are highly common. Remaining constructive stewards over the long term involves more patience and tolerance than many activists would prefer. This is probably why many larger asset managers appear to be more evolutionary than revolutionary in their stewardship practices. Their approach involves a certain appreciation for the uncertainties corporate leaders face in predicting how best to prioritize competing interests among stakeholders, including employees, suppliers, and their communities, and broader environmental concerns which may take decades to manifest.

Investment strategies that intentionally exclude companies that could benefit most from transformation compromise one's ability to shape that transformation and benefit from its success. Because of these inherent risks and weaknesses, investors who genuinely hope to shape change and forge progress may intentionally choose to de-emphasize exclusionary products and pursue more impactful approaches. This does not mean indexed investing is "wrong" in any normative or risk management sense. Such products may effectively protect their owners from certain unwanted risks or facilitate their desires to align their investments with their beliefs, as we have discussed. But screened investments *alone* cannot and will not deliver double-bottom-line returns optimally. Investors who want their investment capital to do well and do good will need to do better. Equally important, more inclusive, sustainable growth would be more effectively promoted if they did.

That said, to lay all responsibility for social progress on investor stewardship teams and publicly listed companies would also be badly misguided. As I have said since this book's outset, we must all fight together for the future we say we want. In so doing, we must recognize that the ability and responsibility of business to change all unwanted environmental and social outcomes is genuinely and appropriately circumscribed. To paraphrase the memorable words most often attributed to Edmund Burke, all that is necessary for the triumph of evil is for right-minded men and women to do nothing. If we all remain satisfied with suboptimal social outcomes, continued environmental degradation, and investment underperformance, such failings will deservedly be our fate. The principal role of the companies we invest in is to allocate their resources mindfully while generating superior long-term growth. The most important business of business is and will always remain the promotion of enduring prosperity, not the single-minded pursuit of cleaner air or enhanced economic mobility. All businesses should also, of course, prioritize environmental and social outcomes when it is in their shareholders' long-run interests to do so. Still, to achieve the many desirable objectives enumerated in chapter 1, much more than heightened mindfulness by business is required. We need individuals to behave differently. We need governmental and regulatory policies to incentivize desired behaviors properly. We also must make sure civic-minded local organizations that excel at delivering proven community-based solutions in real time have the resources they need to succeed.

After all, no community can thrive without self-care by its members.

PART 3
Solutions

Chapter Fifteen
_____

CIVICS LESSONS

In a gentle way, you can shake the world.
MAHATMA GANDHI

The only answer in this life, to the loneliness we are
all bound to feel, is community.
DOROTHY DAY

I don't know about you, but I have heard more than enough about the detailed machinations, promises, and prospects of ESG investing. It's now abundantly clear our existing medley of ESG-integrated investments—worth more than $120 trillion and growing—cannot and will not, on its own, usher in permanent prosperity, a comprehensive solution to climate change, greater economic mobility, or an end to racial injustice. The ESG movement has helped forge more mindful corporate leadership and drawn significant attention to the important roles business can play in improving our societies while generating economic growth. In this sense, ESG has been an unqualified force for greater good. That said, if people somehow believe the ESG movement is on course to solve all of our most pressing problems while simultaneously generating superior returns, on balance, it may well be doing more harm than good. This is why we need to shift our attention away from all the technical issues we've explored—i.e., SASB, TCFD, Article 9, materiality, and other ESG minutiae—and start talking about *proven* solutions. It's time we identify concrete things we know we can do right now to promote more inclusive, sustainable growth. And this is exactly why I must tell you about EMPath.

Since launching their Mobility Mentoring program in 2014, the Boston-based EMPath has helped more than 225,000 human souls achieve greater economic independence and vastly improved living standards. Today, more than 140 member organizations across the United States use some form of EMPath's Mobility Mentoring pedagogy. According to their records, the households they reached in 2020 alone experienced an average annual increase in income of $5,736. Since EMPath's inception, its flagship participants have nearly tripled their annual incomes, from an average of $16,621 to more than $47,000. At the same time, more than 90 percent of their counselees living in temporary shelters moved into permanent housing.

These are stunning results. EMPath has a proven methodology for promoting economic *inclusivity*. They are achieving exactly what we committed ourselves to at this book's outset. They are what I call an *exemplar of hope*. The extraordinary work EMPath has been doing has verifiably promoted economic inclusivity. Hundreds of thousands of human lives have been positively transformed. Their unique methodology can be scaled to help millions of others. And EMPath is hardly alone.

For almost thirty years, the Environmental Defense Fund has partnered with corporations like McDonald's, AT&T, Walmart, and CVS to identify new ways to limit waste, save energy, and derive more of their power from renewable sources. Their Climate Corps and Green Portfolio programs—established in partnership with the private equity firm KKR—have helped firms across China and the United States cut more than 2.2 million metric tons of carbon emissions from their supply chains. These, too, are stunning results. A reduction of 2.2 million metric tons of pollutants means greater economic and environmental sustainability. And the Environmental Defense Fund's success at reducing carbon emissions is expected to produce the same climate benefits as closing one-third of all coal power plants in the world by 2035. Just think about this, briefly. The work of one organization has effectively helped close the equivalent of one-third of the world's coal power plants through greater corporate mindfulness and smart, strategic planning. Isn't that simply awesome?

Like EMPath, the Environmental Defense Fund is using highly effective, scalable methods to promote more *sustainable* economic growth. They are applying proven methodologies and solutions to one of the greatest challenges of our times: getting our greenhouse gas emissions crisis under control. They are also promoting methods that many other corporations can and should follow. They, too, are an exemplar of hope.

Similarly, through their "Learn My Way" and "Make It Click" collection of curated learning tools, the UK-based Good Things Foundation has helped more than 3.5 million individuals who would otherwise have been excluded by the digital divide attain the technical skills they need to access online health, educational, and financial tools. Just like EMPath in the

United States, Good Things partners with thousands of community organizations across the United Kingdom, Australia, and beyond. Their efforts
will ultimately help millions attain digital literacy and connectivity. Make
It Click graduates now see and engage with the world in ways they never
imagined. The Good Things Foundation is yet another exemplar of hope.
Experts and volunteers working with their tools and resources are helping
excluded individuals become more included systematically. They are helping
to close the digital gap between the haves and the have-nots, verifiably, with
an immensely positive human impact.

Though highly disparate in focus, EMPath, the Environmental Defense
Fund, and the Good Things Foundation have three things in common. First,
they are uniquely specialized and highly effective in the important work they
do, arguably best in class. Second, they each in their own way use proven, scalable methods to promote more inclusive, sustainable growth—the tripartite
solution and core objectives to which we committed ourselves at the outset.
Scalability means many millions more can still benefit from their expertise.
Third, each is a nonprofit organization. They are run by neither government
nor business. This nonprofit status confers competitive advantage. Such nongovernmental organizations (NGOs) are uniquely incentivized to be more
concerned about impact than their financial bottom lines. As opposed to the
corporate world, where financial success drives one's potential for impact,
exemplars of hope create impact in ways that drive their financial viability.
Each is also providing benefits to society in ways no commercial enterprise
or governmental program could possibly replicate. The combination of all
these factors means they and other similarly impactful NGOs are part of
what the former chief economist of the IMF and appropriately revered University of Chicago professor Raghuram Rajan calls society's "third pillar."

I will speak a bit more about exemplars of hope and the ways they might
be supported and their efforts scaled with help from investors in my concluding chapter. At the end of this book, you'll also find an appendix with
a representative list of exemplar-of-hope organizations. I encourage you to
submit your own suggestions for inclusion on this growing list at my personal
website, www.1pointsix.com.[1] I believe exemplars of hope can do as much
or more to benefit the world than any public policy or ESG product possibly could. Instead, my intention here is twofold. First, to underscore how
and why nonprofit organizations like EMPath, the Environmental Defense
Fund, and other civic institutions are critical components of healthy nation-
states—something no thriving society can exist without, in fact. Second, to

highlight an especially important subcategory of civic organizations that is now most urgently needed: NGOs that can transform local communities and improve people's lives *at scale*. Essentially, all exemplars of hope are locally engaged, civic-minded NGOs that are proven agents for achieving more inclusive, sustainable growth and stronger communities in proven, systematic manners. The world we say we all want cannot and will not be forged without their successful propagation.

In Professor Rajan's carefully reasoned framework, human flourishing is most commonly found in societies with three highly effective, distinct, yet complimentary pillars: business, government, and community. Without vibrant civic organizations actively muting the excesses of competitive markets and incompetent or overly assertive state agencies, Rajan claims, societies will underperform, or perhaps even rupture. Optimality can be achieved only when all three pillars are performing homeostatically—which is to say, in concert with the others:

> Society suffers when any one of the three pillars weakens or strengthens overly relative to the others. Too weak the markets—and society becomes unproductive; too weak a community, and society trends towards crony capitalism; too weak the state, and society turns fearful and apathetic. Conversely, too much market and society becomes inequitable; too much community and society becomes static; too much state, and society becomes authoritarian. A balance is essential.[2]

Rajan's market–state–community paradigm helps frame possible solutions to the challenges every society faces. It also offers a rough blueprint for more optimal steady-state dynamics that every country should consider. For example, Rajan cites the United States as an example of a modern society that has allowed markets to overwhelm its other two pillars. One can see why he would. The general rise of populism in the United States has been precipitated in part by multiple negative externalities caused by free markets that were left unchecked. After all, unfettered free trade decimated entire communities in the United States—like those in North Carolina that depended on the furniture industry, as well as dozens of cities across the so-called Rust Belt that depended on domestic automobile and industrial manufacturing. At a minimum, some form of worker retraining or relocation services should have been provided by civic organizations or governmental agencies to ease the transition and reposition displaced families. Similarly, a good portion of America's catastrophic opioid crisis can be traced to communities that were mindlessly gutted by free trade, and to unscrupulous pharmaceutical companies that were allowed to run rampant. Markets are highly effective at allocating resources, but all workers need some training, and prudential regulations will always be needed to protect public safety. Multiple externalities threaten desired social cohesion and essential environmental goals. Enlightened government policies or appropriately engaged philanthropic institutions are needed to help restore balance.

Rajan cites China as an example of a society in which government has too much of an upper hand: "China will only continue to grow if it can harness the immense innovative capabilities of its people—as that is the nature of growth at the frontiers. . . . The community will also have to be allowed more freedom and choice, both to sustain innovation but also to maintain healthy separation between the state and markets."[3] To bring the United States, China, and every other nation back into balance, Rajan argues, their weakest pillars need to be reinvigorated. In some cases, this will mean more markets and less government; in others, more civic engagement or better government. The implications are nevertheless clear. If Robert Shiller's good society is to prevail, competent laws and agencies, competitive markets, *and* engaged civic society must all perform at their peaks. No society will thrive unless healthy markets, competent governments, and engaged civic institutions competitively and cooperatively coexist.

A quick tour around the globe further illustrates Professor Rajan's thesis. One sees readily how nondemocratic societies systematically suffer from weak civic institutions. For example, China and Russia rank 126th and 117th, respectively, on the UK-based Charities Aid Foundation's ten-year rankings of charitable giving—that is, at the very bottom of their global survey.[4] The Charities Aid Foundation has measured volunteerism, unconditioned aid, and charitable giving statistics in more than 130 nations over the past decade—a period that includes both financial crises and rapid growth. Russia and China lag in all three categories. These government-heavy states suffer from a lack of civic engagement for an obvious reason: when citizens are told that everything begins and ends with federal authority, self-expression, volunteering, and charitable giving are effectively deemed superfluous. No civic organization can thrive when their governing agents perpetually or repeatedly insist they aren't needed or welcome. The foundation's analysis confirms what Professor Rajan predicated: where governments predominate, community recedes. Other authoritarian European, African, Asian, and Latino countries similarly suffer from weak civil institutions. When governments set behavioral limits and lay claim to all problem-solving, personal and communal responsibility are fatally undermined. Similarly, when markets are presumed to be both omniscient and omni-benevolent, needed interventions from governments and communities are too readily overlooked. Markets will not promote inclusivity or sustainability on their own. Their excesses need to be countervailed, wisely.

While India is a democracy, Rajan notes it still suffers from significant weaknesses in its state and commercial pillars. As we all know, democracy is no sure recipe for government competence. Regrettably, India's democratic government excels at generating rules and red tape. Those rules often hamstring the country's industrial base and compromise administrative effectiveness. Too much or too little government does not equate to good government; only good government does. "India's challenge in the years to come is not its democracy, but the need to strengthen both state capacity and private sector independence," Rajan writes.[5] Because their governments can be so overbearing, Rajan warns, both China and India must also be wary of growing crony capitalism.

Around the globe, philanthropy has followed four broad models: (1) an Anglo-Saxon model, in which civic institutions are largely supported by wealthy individuals to advance causes they consider important, often religious or educational in nature; (2) a Rhine model, in which civic institutions are usually subcontracted by the state to perform "social corporatist" works; (3) a Latin/Mediterranean model, in which the Church is largely seen as responsible for charity and government responsible for broad service provision; and (4) a Scandinavian model. The Scandinavian model relies on a robust public social safety network complemented by established traditions of volunteering. Given that Finland, Norway, Sweden, and Denmark regularly rank among the happiest countries in the world—along with Iceland, Switzerland, the Netherlands, and Bhutan, in one authoritative 2021 survey— something positive needs to be said about the Scandinavian model for forging optimal state, market, and community balance.[6]

When canvassing these four philanthropic models in action, it also quickly becomes clear that strong markets plus charity alone will not be enough for social flourishing. Poverty has persistent intergenerational tendencies. Intergenerational poverty cannot and will not be solved by economic growth and episodic almsgiving alone. Eradicating structural poverty requires structural disruption. Indeed, if almsgiving alone were enough, there would be no poverty in America. By every financial metric, America is the most generous nation on earth.

There are more than 1.5 million charitable organizations in the United States. Over the past two years, Americans have donated nearly $1 trillion to support them. This is by far the largest amount of philanthropic aid of any country, even more than the annual GDPs of Switzerland and Saudi Arabia. Individuals account for 70 percent of these donations, while foundations provided another 17 percent. Bequests and corporate sources compose the remainder. The total volume of financial assets now controlled by U.S. endowments and foundations exceeds $2.5 trillion. This amount is more than the annual GDP of all nations in the world but the top six. It's basically equivalent to France's annual output. Americans are nothing if not generous.

Yet the United States also has the highest level of child poverty in the OECD—double the rate of France and ten times that of Denmark. The United States also ranks near the bottom—33rd of 36—in infant mortality in the OECD. This paradoxical coexistence of exorbitant wealth and persistent want demands more serious introspection. Given the extraordinarily high number of charitable institutions in the United States and their effectively unlimited resources, something other than weakness in the community pillar is needed to explain America's chronic poverty and economic mobility challenges. This paradox demands resolution. The United States has more than enough wealth to eradicate poverty effectively, as many other nations have done. For some reason, it has not done so. Why not? What's missing? Why are there more impoverished children in the United States—by far the wealthiest country in the world—than there are in the rest of the G7 combined?

While American charities and foundations do many wonderful things, as a collective force they would benefit significantly from more rigorous attention to what works and what doesn't. It turns out that more enlightened government programs and more supportive business engagement combined with community initiatives that work are all that is required to forge greater impact and lasting change. Specifically, American governmental policies assisted by the business community are needed to help scale proven civic programs that show the greatest impact, like EMPath's. America's three pillars need to coalesce.

In a landmark study on urban economic mobility in the United States, researchers from Harvard and Stanford found that Salt Lake City has especially beneficial lessons to share.[7] Unsurprisingly, Salt Lake City's relative success relates in no small part to the prevalence of Mormon traditions and the principle of self-reliance upon which the Church of Jesus Christ of Latter-day

Saints (LDS) largely depends. The LDS welfare manual explicitly extols the importance of work: "The Lord has commanded us to work, for work is the source of happiness, self-esteem, and prosperity. It is the way we accomplish good things in our lives."[8] As with Catholicism and other religious traditions, LDS leaders also insist that responsibility for all well-being resides first with the individual. But LDS does not stop at self-reliance; they cannot since self-reliance is not sufficient. When individuals struggle in LDS communities, the second line of defense must be family. And when family falters, the third line of defense is the Church. As it so happens, family and the Church are very powerful forces for social cohesion in Salt Lake City. In the LDS tradition, government welfare is not called upon until individual, familial, and Church resources have been exhausted. If you are part of a Mormon parish (as many are in Utah) and something is not going well in your life, more likely than not members of your local parish will intercede, helping you get back on track. The Harvard and Stanford researchers found communal responsibility like this to be especially effective. Because of political consensus in Utah, more-over, LDS community traditions have been supported and further ampli-fied by government agencies. Utah's Intergenerational Poverty Mitigation Act requires all state agencies to share administrative data on their clientele, meaning their programs are adept at keeping track of all Utahns, especially at-risk residents. Community cohesion combined with supportive public policy translates into fewer folks falling between the cracks.[9]

The Wilson Sheehan Lab for Economic Opportunities (LEO) at the Uni-versity of Notre Dame is one of several leading programs in the United States that rigorously applies evidenced-based scientific methods in their mission-driven quest to identify and scale the most effective solutions to persistent social problems. From education and health projects to housing, criminal justice, and methods to achieve economic self-sufficiency, LEO partners with NGOs and government agencies across the United States in pursuit of a singular objective: identifying innovative policies and programs that verifi-ably benefit those in need and promote broader human flourishing. In short, LEO's research is dedicated to supporting today's and discovering tomor-row's exemplars of hope.

An example of the type of program LEO evaluates is the Nurse-Family Partnership (NFP). NFP began as a small home visitation program for low-income new mothers in Elmira, New York, a small town with a population of only twenty-eight thousand. Several randomized controlled trials veri-fied that NFP's approach of providing regular home visits by trained medical professionals to young, at-risk mothers dramatically improved health out-comes for both mothers and children. Moreover, these trials found that every dollar spent on NFP services saved $5.70 in future state costs, with the highest-risk families benefiting most. Armed with this data, NFP was able to apply for and receive government support to scale beyond Elmira's town lines. Today, more than 340,000 families in forty states and multiple indigenous communities have been NFP beneficiaries. NFP methodologies are helping young single moms and their kids stay healthy in a cost-effective manner, reducing child neglect by 48 percent and behavioral problems by 67 percent. In short, Elmira's successful local innovation has gone viral. As you would expect, NFP is an exemplar of hope. They are another highly effec-tive, evidenced-based program that can be scaled through government and business partnerships, dramatically amplifying more inclusive, sustainable growth. Identifying programs like NFP and EMPath is LEO's raison d'être.

To this end, LEO recently examined the success rates of community col-leges. Although 70 percent of U.S. high school graduates head off to some type of college, only about 35 percent end up with a degree, a disappointing 50 percent completion rate. After examining the successful experience of Catholic Charities Fort Worth's Stay the Course program, LEO generated analysis that has since led to the implementation of a new national policy proposal. Data showed that at-risk students who were assigned one-on-one community college guidance counselors provided by Catholic Charities Fort Worth achieved completion rates 400 percent higher than those with-out mentors. It turns out money was not their problem; developing imple-mentable plans to juggle young parenthood, find affordable day care, balance course loads, and arrange reliable transportation to and from class was. Struggling community college students needed advice about logistics, not additional economic incentives. Given that more than half of all jobs in the state of Texas rely upon some form of professional attainment, which com-munity college programs provide, Stay the Course is now positioned to help thousands more aspiring, at-risk Texans attain self-sufficiency. The Institute for College Access and Success has also taken this program to communities across California and Michigan, scaling its proven method. What we should now ask ourselves is this: If Stay the Course is working in Texas, California, and Michigan, how could we scale it to other states so they too benefit from its methods?

Achieving more inclusive, sustainable growth systematically in the United States and around the world will require active engagement within and between markets, government, and community—that is, Professor Rajan's three pillars. When it comes to the specific contributions that need to be made by community, more support for evidence-based, scalable solutions like Make It Click and Stay the Course should become societal priorities. Exemplars of hope have proven programs mission-ready to be scaled. All we need to do is scale them.

Human flourishing does not transpire at a state or global level. States are political constructs, not human ones. Human flourishing exists only within the context of real, identifiable communities. Countries ultimately succeed through the aggregation of flourishing communities. The world will succeed only through the broad aggregation of flourishing communities, too, aided by enlightened leadership. As Catholic activist and canonization candidate Dorothy Day witnessed and lived, everything begins and ends with one's community. And the strongest models for identifying and scaling solutions to help communities flourish are what the American Ideas Institute chairman, Jeremy Beer, has memorably called "philanthrolocalism."

According to Beer, the primary purpose of philanthropy should be to increase the opportunities for and strengthen the possibilities of authentic human community: "The forces of modern life have conspired to fragment and weaken many, if not most local communities. Industrialization, globalization, mass culture, modern warfare and geographic mobility, to name a few factors, have enriched the lives of a few fortunate places while depleting the vitality of many, many others."[10] In Beer's telling, giving locally, focusing on local solutions, and serving one's immediate neighbor rather than "changing the world" are what will help the human family most. Localism would be a deep mindset change for many corporations and philanthropic institutions; more often than not, they want to "change the world." But such local engagements are the best, if not the only, way to redress the urgent needs of real people systematically.

Through the widespread fortification of individual communities, more inclusive growth could be achieved. Through the active propagation of best environmental practices, more sustainable growth could be achieved as well. We need the Environmental Defense Fund to consult with every major corporation, not just those which can afford their services. We need Make It Click to become common training across thousands of communities in every country, not just a chosen few. We also need NFP programs to be more widely available in communities where such support is most needed. In my opening chapter, I identified many of the problems that need to be fixed.

What we need now is a focused, collective mindset and dedicated resources to scale proven programs that verifiably work.

Consider, for example, successful businessman George B. Kaiser. He has taken the concept of philanthrolocalism to a paradigmatic level in his native city of Tulsa, Oklahoma. "No child is responsible for the circumstances of his or her birth," Kaiser rightly observes. "I had the advantage of both genetics and upbringing. As I looked around at those who did not have these advantages, it became clear to me that I had a moral obligation to direct my resources to help repair that inequity." With generous support from the Kaiser Family Foundation, children now born to disadvantaged families in and around Tulsa—a community of about one million people—are invited to participate in the foundation's Birth Through Eight Strategy for Tulsa (BEST) program. BEST consists of a network of multiple complementary programs and services, some public, others private. Together they help ensure all young Tulsans receive natal health care, positive developmental education, and academic success by third grade—all factors that have been proven essential to success later in life. At this writing, BEST has impacted 80 percent of its target population and dramatically improved outcomes for thousands of children in their earliest years of development. Clearly, George Kaiser is a devoted family man—it's just that every child born in and around Tulsa is an integral member of his extended family.

Exemplars of hope are organizations that have scalable methodologies that meaningfully improve the lives of many—organizations like the Nurse-Family Partnership, Catholic Charities Fort Worth, and the Kaiser Family Foundation. Of course, organizations like these cannot and will not cure all environmental and societal ills on their own. We still need complementary, farsighted, and supportive public policies. Critically, we also need thriving commerce. Human flourishing requires proficiency in all three pillars. Exemplars of hope are essential contributors to community proficiency. Exemplars of hope are what make community pillars strong and effective.

We have now come to our final chapters on the optimal roles business and finance can play in fashioning Robert Shiller's good society. As we pull all the history and arguments offered here together, we will draw upon the important insights of Professor Rajan and Jeremy Beer, as well as our prior discussions on indexed investing, ESG methodologies, and promising frameworks, like the Shared Value Initiative. Societies flourish when markets, communities, and the state each perfect their comparative advantages and perform complementary roles. Thriving communities need successful individuals to help care for and nurture their less successful members. Thriving communities need vibrant economies. And thriving communities need effective, enlightened governments. Philanthropic efforts that effectively address local needs are essential for nurturing community. Proven environmental and social programs that can be efficiently scaled from one community to another are indispensable for systemic improvement. Indeed, widespread, inclusive, sustainable growth will not be possible without their broader propagation.

But where exactly does all this leave finance and ESG investing? In my considered opinion, it both opens and refines ESG's aperture for optimal impact. As you are about to see, the world has more than enough money to solve all of its most pressing social and environmental ills. To optimize societal and environmental outcomes, our collective financial resources need to be more mindfully allocated. In particular, assets under the direct control of high-net-worth individuals and institutions like pension plans, insurance companies, sovereign wealth funds, and endowments need to be modestly recalibrated to achieve more tangible and measurable impact. Casey Quirk imagined a future world where a good portion of ESG investments would migrate from simple risk-integrated strategies to those with more impactful outcomes, that is, where tangible environmental and social objectives are achieved. If finance is to help increase the probability of more inclusive, sustainable growth, this reallocation needs to materialize. We have already seen that green, social, and sustainable impact bonds are examples of ESG investments with verifiable impact. But they are only a small part of the exciting broader opportunity set that impact investing now presents. When combined with impact venture capital, impact public and private equity, and impact real asset strategies, Archimedes's lever and fulcrum begin to come into plain view. Most importantly, Shiller's depiction of investment managers who "are among the most important stewards of our wealth, and vitally important players in the service of healthy and prosperous market democracies" is given a road map for moving from theory to practice. And if all this happens in the years immediately ahead, the greatest chapters of human flourishing will be the gift we can leave to generations unborn.

Chapter Sixteen

IMPACT INVESTING AT SCALE

I can see that a solution is within our grasp; I call it the
Impact Revolution. . . . It will allow us to address the
dangerous inequality and degradation of our planet.
It will lead us to a new and better world.

SIR RONALD COHEN

What would you say if I told you there are many ways to promote more inclusive, sustainable growth *and* make a market return? What if I were to add that you wouldn't have to guess about the environmental and social impacts of your investments, like you must do with most other investments, including many indexed ESG products? Rather, you would know about your impact specifically and verifiably, with visual attribution reports that might look like the one in figure 16.1.

What does this visualization depict? It shows that the companies in which Baillie Gifford's Positive Change public equity team invested in 2020 helped deliver multiple objectives sought by the UN's Sustainable Development Goals, often very meaningfully. During 2020, the companies owned in Baillie Gifford's primary impact investment fund (1) delivered mobile and digital services to at least one billion people; (2) enabled customers to save a total of 1.2 trillion liters of water; (3) protected 540 million people from communicable disease; (4) provided more than 1.5 billion people access to financial services; (5) facilitated nearly eleven million telehealth visits; (6) helped consumers eliminate seventy million metric tons of atmospheric CO_2 (the equivalent of removing fifteen million passenger cars from the road); and (7) improved environmental farming practices on 233 million hectares of land, an area nearly ten times the size of the United Kingdom. These same companies also prevented more than 260,000 metric tons of waste from being created.

FIGURE 16.1. Baillie Gifford impact report.
Reprinted with permission from Baillie Gifford's Positive Change Impact Report – 2020.

This is a lot of tangible impact. As far as promoting inclusive, sustainable growth, it's hard to imagine a selection of several dozen publicly listed companies doing more. Oh, and did I mention that this portfolio produced an annualized return of 43.64 percent from January 2020 through December 2021? Over that same two-year period, their MSCI benchmark returned just 17.39 percent per year. That means Baillie Gifford's Positive Change Team produced 26.25 percent of *positive* tracking error, or alpha, per year from 2019 through 2021. In short, they massively outperformed their stated goals: (1) to deliver attractive long-term investment returns of at least 2 percent above the MSCI ACWI over the investment cycle; and (2) to contribute meaningfully toward a more inclusive and sustainable world.

My word! How's that for doing well and doing good?[1]

Impact investing at its best generates verifiable social and environmental progress. That progress is tangible, measured, and often expressed in terms relating to the UN's Sustainable Development Goals. Most impact investing strategies simultaneously generate a financial return. Sometimes that financial return matches or exceeds market indices; sometimes it doesn't. Impact investment strategies exceeding market returns effectively means that the beneficiaries of those strategies are doing very well and a lot of good at the same time. Of course, this is what many mindful investors hope to do— generate a double bottom line. Successful impact investing demonstrably promotes Shiller's good society.

Figure 16.2 provides a visual depiction of how various ESG equity investment strategies perform in the known return-versus-impact space. This figure depicts how different types of ESG equity investment strategies perform relative to their chosen indices and the amount of *knowable* impact they provide. Most ESG equity indexed strategies have very low tracking error combined with low levels of documented, verifiable impact. They are broadly risk mitigating and return seeking. In contrast, active ESG equity strategies like the Funk and Powell fund at Brown Advisory or the activist fund at Inclusive Capital Partners launched by Jeffrey Ubben are likely to have more knowable social impact along with wider tracking error. A third category, known as ESG impact strategies, has even wider tracking error along with more verifiable impact. Verifiable impact is the sine qua non of all impact investing, in short. ESG impact strategies stand out from other ESG strategies because they directly promote specific social and environmental objectives that can be traced to the UN SDGs in measurable ways while providing quantifiable financial returns. With most impact investment strategies, you receive assurance that you are doing something good because you are given detailed, audited impact reports that show what your investments did and how they did it.[2]

Public equity strategies aren't the only ones striving for impact, moreover. A number of public debt, private equity, private debt, and many real-asset investment strategies also include explicit impact objectives.

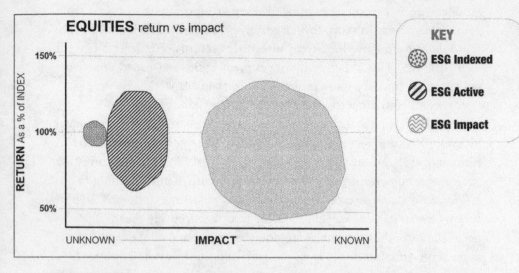

FIGURE 16.2. Public and private ESG equity strategies: return versus impact.
Figure by the author.

Consider Bain Capital's Double Impact fund. It is a private equity strategy largely focused on the United States. It targets three primary opportunity sets: health and wellness, education and workforce development, and sustainability. Bain's overarching impact strategy is to scale mission-aligned companies in ways that deliver both competitive financial returns and meaningful social and environmental improvement. As active owners, they single out companies capable of making transformational change and then resource them to realize their full potential. The impact team at Bain Capital also frequently partners with each of their companies to improve their B-Corp scores. One of their companies, Rodeo Dental, became the first dental service organization to win full B-Corp accreditation. An example of a corporation that Bain Capital purchased, helped transform, and then exited is Penn Foster, a leading digital learning platform. While under Bain's control, Penn Foster's earnings before interest payments and taxes more than doubled as their educational programs completion rate grew 35 percent and course enrollments grew to a record sixty-four thousand. This is inclusive, sustainable growth in action. Bain Capital's first impact fund, which included the Penn Foster investment, is reported to have achieved annualized returns of approximately 25 percent per year. While this would be below other private equity strategies over the same period, it still exceeded the performance of many public stock indices, meaning investors were fairly compensated for the illiquidity they incurred. In short, it did well and did good. A second Bain Capital impact fund of nearly $800 million closed in 2021. As with their first vintage, their second Double Impact fund buys companies with promising fundamentals and proven social or environmental credentials. The Bain Capital team then helps turn them into higher-performing, more valuable firms.

Another market leader in the private equity impact space is San Francisco–based TPG. After raising more than $6 billion in 2021 for a new climate-focused fund for which the ex-U.S. secretary of the treasury Hank Paulson serves as the executive chairman, TPG's family of so-called Rise Funds now includes more than $12 billion in client capital. While TPG's Rise Climate Fund takes a broad-sector, environmental approach—ranging from growth equity to value-added infrastructure—other funds in TPG's Rise family focus on growth opportunities in health care, education, and economic mobility. TPG's impact investing funds have a proven record of working with early-growth-stage, high-potential, mission-driven companies that have the power to change the world.

Similarly, the impact investment team at Apollo—another large alternative asset manager—believes private enterprise has an important role to play in improving the world, both socially and environmentally. They use the power of their franchise and a differentiated strategy to promote impact at scale by investing with a high degree of intentionality in later-stage businesses that are helping people or healing the planet. One such example is Reno De Medici (RDM). RDM is a leading recycled-carton-board manufacturer with a pan-European asset base, a large market share, and the number-one position in its core markets. RDM plays a critical role in the circular economy by recycling wastepaper into new packaging material, helping to make the packaging industry more sustainable. RDM's financial performance is inherently linked to this positive environmental impact by virtue of its business model, a characteristic known as "collinearity." As RDM's sales volumes grow, substituting less sustainable alternatives like plastic and virgin fiber packaging, the company's revenues grow. As RDM reduces production-related carbon emissions, water use, and waste generation, its margins grow owing to reduced associated expenses in its cost structure. Collinearity in a company's business model gives Apollo confidence that impact and financial performance can grow in tandem and be mutually reinforcing. Additionally, RDM, like all of Apollo's impact investments, is given specific goals to improve its B Impact Assessment, a generally accepted industry standard of impact, in addition to financial metric goals. Apollo sees multiple opportunities like RDM that can drive impact at scale through mature businesses, while also delivering attractive, nonconcessionary returns for their investors.

While Bain's private equity impact funds focus almost exclusively on the United States, those of LeapFrog Investments—founded in 2007 by Andrew Kuper, a renowned entrepreneur who helped pioneer "profit with purpose" investing—focus exclusively on emerging markets. Today, LeapFrog owns significant stakes in high-growth health care and financial services firms across much of Africa and Asia. LeapFrog also consulted on the creation of the Operating Principles for Impact Management,[3] helping to develop voluntary investment standards for the entire field. LeapFrog secured capital from a number of sophisticated institutional investors, including the Singapore-based Temasek and Zurich Insurance. Their board also includes a former prime minister of Australia. LeapFrog focuses on five of the UN's SDGs: SDG 1, no poverty; SDG 3, good health and well-being; SDG 5, gender equality; SDG 8, decent work and economic growth; and SDG 10, reduce inequalities. In 2020, the companies in which LeapFrog invested provided health care and financial tools to 221 million individuals, 80 percent of whom were in low-income emerging markets, typically living on less than $10 per day. Over that same period, the value of their portfolios grew 22 percent. For LeapFrog, "profit with purpose" isn't just a motto; it's their raison d'être and core to everything they do.

India-based Dvara Holdings is an example of the type of company in which LeapFrog and other impact investment firms may choose to invest. Dvara provides millions of low-income households in India with microloans, small business finance, and other forms of credit. As COVID-19 exacted tolls on many of Dvara's clients, LeapFrog helped Dvara increase its digital collection and disbursement capabilities, reducing operational costs and increasing the efficiency of their operations significantly. For many of Dvara's clients, digitization meant survival. For LeapFrog and their investors, it meant higher returns.

In the real-asset market—a broad field encompassing multiple forms of commercial and multifamily residential housing, as well as infrastructure like power plants and logistics facilities—a growing number of green projects are directly lowering carbon emissions while generating at- or above-market rates of return. Three megatrends are driving the sustainable opportunity set in infrastructure: (1) decarbonization; (2) digitalization; and (3) decentralization. *Carbonomics*—a research initiative led by Goldman Sachs—projects at least $50 trillion will be put to work by 2050 in pursuit of net-zero goals.

Multiple green infrastructure funds have generated historic gross returns of 12 to 14 percent per year, with 7 percent average cash yields. These funds focus on carbon capture and storage, renewable energy power production, and industrial process improvements. Their returns are uncorrelated to the broader equity market. They also provide inflation protection that other asset classes do not. Inflation protection and noncorrelation are valuable attributes for portfolios seeking greater diversification and lower overall volatility. Optimal portfolios cannot be built without them.

Blackstone—the world's largest alternative asset manager—has developed a climate program to help portfolio companies decarbonize in order to become more valuable. Blackstone works closely with its portfolio companies, drawing on the firm's expertise in reducing carbon emissions across a wide variety of assets in vastly different sectors of the economy. The firm has committed to a goal of reducing carbon emissions by 15 percent in aggregate for all new investments for which it controls energy use over the first three years of ownership—a target that is guided by climate science. In Blackstone's privatization of the Hilton Hotels & Resorts group, Hilton was able to cut carbon emissions by 30 percent, waste by 32 percent, and energy and water usage by 22 percent before returning to public ownership at a multiple of its purchase price. Greater attention to their environmental footprint saved Hilton more than $1 billion in operating costs. Value creation through environmental stewardship is not confined to public corporations alone, in other words; given the close partnership and long-term investment horizon, significant value can be created for real assets through sustainability as well. Blackstone is also undertaking a carbon-accounting program that will allow the firm to measure its progress. In sum, Blackstone is striving to play a significant role in building a stronger economy by helping to finance global decarbonization. Since 2019, the firm has committed nearly $10 billion in investments that are consistent with the broader energy transition.

One of the more important areas of investment impact that has not only a proven track record but also considerable unmet investment need is renewable power. While significant efforts are underway to electrify more everyday activities—from transportation to home heating—these conscientious efforts won't matter much for the environment unless the sources of electricity that power them are also derived from more sustainable processes. According to John Doerr, to have any hope of reducing our carbon emissions dramatically, "every fossil fuel power plant will need to be shuttered."[4] In emerging markets alone, it is estimated that $1.5 trillion of incremental investments are needed per year to phase out coal while meeting growing energy needs. The vast majority of this funding needs to come from private sources.[5] In recent years sovereign wealth funds in particular have devoted about $80 billion to renewable energy projects globally—helpful to be sure, but only a fraction of what is actually needed (figure 16.3).

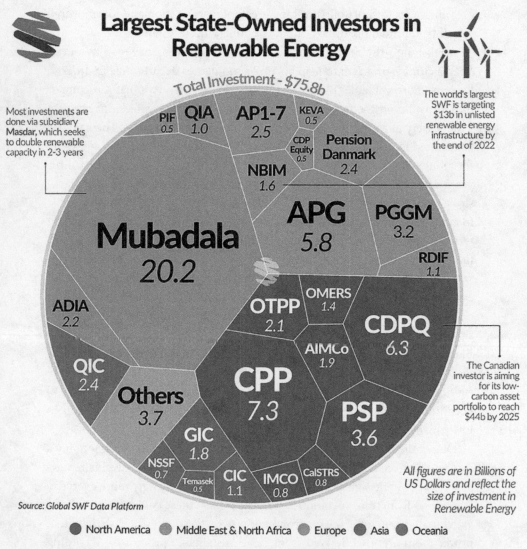

Largest State-Owned Investors in Renewable Energy

Total Investment - $75.8b

Most investments are done via subsidiary **Masdar,** which seeks to double renewable capacity in 2-3 years

The world's largest SWF is targeting $13b in unlisted renewable energy infrastructure by the end of 2022

PIF 0.5 | QIA 1.0 | AP1-7 2.5 | KEVA 0.5
CDP Equity 0.5 | Pension Danmark 2.4
NBIM 1.6
Mubadala 20.2
APG 5.8 | PGGM 3.2
RDIF 1.1
ADIA 2.2
OTPP 2.1 | OMERS 1.4 | CDPQ 6.3
QIC 2.4
AIMCo 1.9
Others 3.7
CPP 7.3
PSP 3.6
GIC 1.8
NSSF 0.7
Temasek 0.5 | CIC 1.1 | IMCO 0.8 | CalSTRS 0.8

The Canadian investor is aiming for its low-carbon asset portfolio to reach $44b by 2025

All figures are in Billions of US Dollars and reflect the size of investment in Renewable Energy

Source: Global SWF Data Platform

● North America ● Middle East & North Africa ● Europe ● Asia ● Oceania

FIGURE 16.3. Decarbonizing emerging-market energy grids is essential for global sustainability. According to multiple industry estimates, an additional $1 trillion per year is needed from public and private sources to achieve net zero carbon emissions in emerging markets.
Reprinted with permission from Global SWF Data Platform.

Impact investments span from New York City, where low-income teen-age mothers may take up residence in apartments that have been built via concessional loans, to Cambodia, where microfinance allows small business owners to stock more needed goods or train new workers. The Global Impact Investing Network (GIIN)—an NGO dedicated to increasing the scale and effectiveness of impact investing—estimates that about $715 billion had been invested in impact strategies by the end of 2020. GIIN surveyed the historic performance of these strategies across both developed and emerging markets, including private equity, private credit, and real assets. Somewhat surprisingly, they found very little difference in the financial performance of developed versus emerging markets. In fact, positive annualized, real returns were achieved even in those strategies based on concessional terms. Private credit strategies produced average annual returns of 7 to 10 percent since 2011, while average private equity returns ranged from 10 to 18 percent over the same period. Only real-asset strategies produced substantively different average returns between developed and emerging markets: 13 percent vs. 8 percent, respectively. Figure 16.4 illustrates the range of returns for each strategy. They are based upon one standard deviation move in each asset

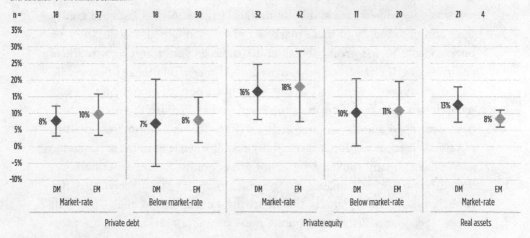

Average realized gross returns since inception for private markets investments

Number of respondents shown above each bar; year of first impact investment ranges from 1956 – 2019, with 2011 as the median year. Averages shown beside each diamond; error bars show +/- one standard deviation.

Source: GIIN, 2020 Annual Impact Investor Survey

FIGURE 16.4. A number of private debt, private equity, and real-asset impact strategies are delivering market-related returns while simultaneously delivering tangible environmental and social benefits.

Reprinted with permission from Global Impact Investing Network, "Annual Impact Investor Survey 2020."

class. The ranges show that there is wide variability in the performance of alternative asset-class impact strategies around the globe but that attractive returns are broadly available to those investors who do proper due diligence and careful manager selection. This is nothing unique; due diligence and mindful manager selection are essential for choosing any top investment product or strategy, whether it is impact seeking or not.

The public and private equity and real-asset markets have a growing number of opportunistic strategies that may reward investors who want their money to do well and do good. Dozens of ESG investment strategies that provide verified and verifiable impact reports in the equity, venture capital, real estate, and infrastructure markets already enable investors to earn market-related returns on their invested capital while simultaneously promoting specific social and environmental goals. A select number of these impact investment strategies are at the efficiency frontier, meaning they provide optimum amounts of impact relative to their returns or optimum returns relative to their impact, all along a continuum. That said, when it comes to turbocharging more inclusive, sustainable growth, impact equity and alternative asset-class products pale in both dollar amounts and potential significance to the debt and loan markets.

In the past few years alone, trillions of incremental dollars have been raised by federal and local governments, supranational agencies, financial institutions, and corporations in the green, social, sustainable, and sustainability-linked loan and bond markets (figure 16.5). If issuance guidelines and the sanctity of these markets can be maintained, it would not be unreasonable to see green, social, and sustainable debt and loan markets continue to grow at high double-digit annual rates. This means that public debt markets could ultimately channel tens of trillions of incremental investment dollars to many of our most urgent social and environmental needs in both developed and developing markets alike. This is an immense amount of capital. There is no doubt it could drive an immense amount of impact.

We have already spoken a bit about green bonds. Unlike sustainability-linked bonds—which finance the *general* functions of an issuer that has explicit sustainability targets linked to the conditions of that bond—all green, social, and sustainable bonds must promote *specific*, tangible projects and outcomes. For example, proceeds from green bonds can be used to finance projects ranging from renewable energy plants, cleaner transportation systems, and wastewater management to improved energy efficiency, climate change adaptation, and greener buildings. The proceeds of social

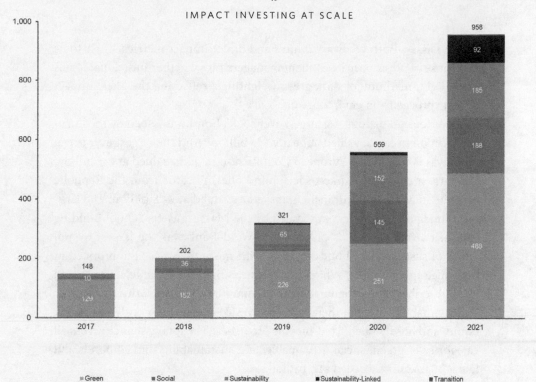

FIGURE 16.5. **The green, social, sustainable, and sustainably-linked bond and loan markets are channeling trillions of dollars of incremental investment to projects that promote more inclusive, sustainable growth.**
Citibank Sustainable Debt Issuance, January 2022.

bonds are equally wide-ranging. They can be used to promote food security, reduce recidivism, create affordable housing, improve access to essential services, and attain improved health objectives, including many objectives relating to COVID-19 concerns, among many other goals. Proceeds from sustainability bonds can be used to finance or refinance a combination of green or social projects or activities. Guidelines for all these markets are set by the International Capital Market Association (ICMA). ICMA green, social, sustainable, and sustainability-linked bond principles are helping to promote transparency and consistency. These principles are as important as the emerging SASB and TCFD standards, if not more. ICMA guidelines are voluntarily followed by issuers and investors alike. In practice, third-party advisory firms like Moody's and Sustainalytics may also rate individual issues for their levels of verified impact. Private managers like PIMCO and NN Investment Partners may also apply their own models and evaluation tools to assess both creditworthiness and direct impact metrics. Their tools and those of other peer investment managers rate whether individual issues merit light, medium, or dark green weightings, reflecting the likely impact of their proceeds on environmental goals.

As of the end of 2021, the largest ever social bond was issued by the European Union in 2020, valued at nearly €17 billion. And the largest ever green bond was issued by the European Commission in 2021, valued at €12 billion. The largest *emerging markets* social bond offering came from the Republic of Chile in 2021, a multitranche transaction valued at $5.8 billion. The largest municipal social bond was issued by the Massachusetts School Building Authority, valued at $1.45 billion. The World Bank was the largest overall issuer of sustainability bonds in 2020, with more than $54 billion priced in fifty-one separate deals. While these volumes may sound large, all were well oversubscribed and sizable enough to enjoy ongoing liquidity in the secondary market. In sum, there is keen investor demand for truly impactful, liquid, and highly rated fixed-income securities. The total issuance across all categories of green, social, sustainable, and sustainability-linked bonds and loans in 2020/21 exceeded $1.5 trillion.

One of the more innovative social bonds issued in recent years came from the Ford Foundation in 2020. It set a new standard for a private foundation committed to promoting more inclusive, sustainable growth. Rated triple A and coming in two tranches—with thirty- and fifty-year maturities—the Ford Foundation raised nearly $1 billion at a combined cost of less than 2.75 percent per annum, which is quite low by historic standards. The proceeds of this issuance were used to support other philanthropies that were struggling with financing while the COVID-19 pandemic became more acute. It is important to understand how elegant this transaction truly was. The Ford Foundation avoided selling securities in their portfolio at depressed prices caused by COVID concerns, while amplifying their disbursements at a time of growing need. Grant recipients included the Environmental Defense Fund, Oxfam, and Enterprise Community Partners, i.e., many exemplars of hope. The latter supports community development and upward mobility with a focus on affordable housing. The Ford Foundation responded to acute and growing need in a timely and impactful fashion by using the social bond market. They did so in a financially responsible manner. Proponents of more inclusive, sustainable growth should be pleased they did.

Like figure 16.2, figure 16.6 depicts the profiles of regular investment-grade debt; mixed portfolios of green, social, sustainable, and sustainability-linked

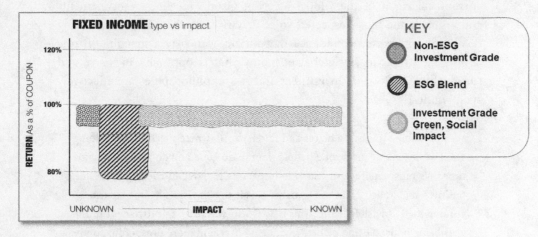

FIGURE 16.6. Investment-grade fixed-income securities in the return-versus-impact space.
Figure by the author.

securities; and pure versions of the latter in the return-versus-impact space. In this case, return is expressed as a percent of coupon set at the time of pricing. As default rates are extremely low in investment-grade securities, the likelihood of getting one's principal and coupon back is quite high, as figure 16.6 depicts. Fixed-income ESG blend strategies involve securities that may not be investment grade–rated. As these have higher risks of default, their returns may be less than the coupons at issue. The full return of one's original principal is also less secure than with investment-grade strategies. Figure 16.6 portrays in stylized form much the same information as figure 16.2, but for bonds rather than equities. Like figure 16.2, its conclusions are equally clear: if investors want to do well and do verifiable good in the fixed-income market—i.e., to achieve market-based returns and verifiable impact—those objectives are more likely to be achieved in the green, social, and sustainability categories rather than in plain vanilla investment-grade bonds.

This brief, albeit comprehensive overview of the impact investment market—a young yet rapidly growing collection of investment strategies that includes the public and private debt and equity markets, as well as real-asset markets like infrastructure and real estate—illustrates the broadening opportunity set for investors who want their money to produce market-related financial returns as well as tangible social or environmental benefits. Its increasing depth and sophistication have given Sir Ronald Cohen—the current chairman of the Global Steering Group for Impact Investment[6] and arguably the most respected voice driving the global impact investing revolution—increased confidence that finance can play a meaningful role in forging an expanding global economy that is optimally inclusive and sustainable. Sir Ronald's investment and moral philosophies are effectively synonymous:

> We are moving towards a better and fairer world, where markets drive doing good while making profit, and people want to do good and do well at the same time. We must embrace measurable impact as a driver in every investment, business, and policy decision we make. This is the "invisible heart of markets," guiding their "invisible hand." This new world will drive improvement in the well-being of people and planet, creating a fairer and more prosperous future for us all.[7]

As someone who has spent four decades of his professional career advising pension plans and sovereign wealth funds, central bank reserve managers and insurance companies, endowments and family foundations, hedge funds and asset managers, and—as importantly—my own brothers, sisters, nieces, and nephews on how to invest their savings to achieve their financial objectives, I do not believe that "every investment" should include measurable impact as a determinant factor. In fact, most portfolios should optimize risk and reward first, with impact further behind as a priority. For example, "tier-one capital" and fiduciary rule regulations about which we spoke earlier largely preclude the elevation of impact over other considerations like liquidity or return for many types of investors and jurisdictions. It is implausible to think "impact" should be a goal of every capital allocation. Consider central bank reserve managers, vital stewards of one key element of every country's sovereign wealth. Central bank reserves amounting to $13 trillion need to be managed for liquidity and safety first, not return and certainly not impact. After all, foreign exchange reserve assets are a country's first line of defense during financial crises. Central banks must retain reserve assets that are immediately deployable under the most exigent of market circumstances, including market crashes or global military conflicts. In practice, this means holding lots of government bonds and bills, not higher-risk asset classes or impact investments. Similarly, no retiree living off their limited savings and the modest annual income those savings generate should prioritize social impact over their own daily needs. Predictable, safe, reliable returns are much more important to retirees living on fixed incomes. Investing for impact shouldn't be their most defining priority.

But now comes some good news. It turns out that "every investment, business, and policy decision" individuals and institutions make need not be impact driven for the UN's Sustainable Development Goals to be achieved. Moreover, it turns out that the emissions goals of the Paris Agreement are not beyond our reach financially, nor is the dream of eliminating extreme poverty or providing quality education to every human soul that is yet to come into the world. In fact, to create a world of optimally inclusive, sustainable growth, all investors would need to do is pledge a small portion of their assets to impact outcomes each year for the next decade. If every institutional and high-net-worth investor followed this very modest asset reallocation, moreover, the global economy would be able to expand more inclusively and sustainably. In other words, all could benefit.

How small, you wonder? Well, how does 1.6 percent per year—or less than two cents for every dollar of savings—sound to you?

Chapter Seventeen

THE 1.6 PERCENT "SOLUTION"

If the human person is not at the center, then something else gets put there, and that's what all human beings then have to serve.
POPE FRANCIS

Capitalism isn't immoral: it's amoral. It's a wild beast that needs to be led.
BONO

If you are at all like me, you recoil when someone tells you they have a simple answer to an extremely difficult, if not to say intractable, problem. To cure income inequality, we need only improve public education. Peace in the Middle East will be possible once combatants learn to love their children more than they hate those on the other side. To end juvenile delinquency, nuclear families must be held intact. Claims like these are not untrue, of course. Indeed, each of these "solutions" would prove highly accretive. Though completely factual, they're factually incomplete. By themselves, they would prove insufficient for resolving the complex problems they target.

This is why I have intentionally put "solution" in quotation marks. Fashioning a world of optimal inclusiveness and sustainability while continuing to maintain the economic growth we need requires comprehensive changes to public policies, corporate mindfulness, investment stewardship, consumption patterns of billions of people, and many other measures. Each of these changes would be revolutionary, and each must be sustained over time. No one should pretend any comprehensive solution to our environmental, social, and economic challenges will be unidimensional or effortless. At another level, however, achieving our many environmental and social goals involves pure and simple math. To promote ubiquitous, sustainable human flourishing for generations to come, we will need to mobilize trillions of dollars of incremental, targeted investment, beginning immediately. When I say "solution," all I really mean is that we can make that math work. As ambitious as it sounds, it is possible to find the trillions of dollars we need to achieve the UN's Sustainable Development Goals, as well as other priorities. It is also possible to direct those incremental trillions to the outcomes that civic-minded individuals and the broader public say they want.

In his *Roadmap for Financing the 2030 Agenda for Sustainable Development*, Secretary-General António Guterres details three objectives, six action areas, and fifteen asks. His objectives include (1) aligning global economic policies and financial systems with the UN's multifaceted 2030 agenda; (2) enhancing sustainable financing strategies and investments at both the regional and country levels; and (3) seizing the potential of financial innovations, new technologies, and digitalization to provide more equitable access to finance. Guterres's UN 2030 agenda maintains alignment with the goal of net zero carbon emissions by 2050. His plans also carry an estimated annual price tag of about $2.5 trillion for emerging countries alone.[1] Another more recent report by the BlackRock Investment Institute estimated that at least $1 trillion of additional, targeted energy transition investments per year are needed.[2] If you net the figures in these reports, subtract likely funding to come from public sources, and add additional environmental and social spending through public–private partnerships that is now needed in developed markets over the next decade, you arrive at an estimated, incremental annual investment price tag of $3.5 trillion per year.[3]

Importantly, this incremental $3.5 trillion per year must come from private sources; public coffers and most governments are already tapped out. While $3.5 trillion seems like a lot, this figure needs to be put into perspective. The total volume of financial assets under the direct control of ultra-high-net-worth individuals and global institutions—including sovereign wealth funds, pension plans, insurance companies, and central banks—is now estimated to be in excess of $220 trillion.[4] To meet the most urgent challenges of our time, this particular group of asset owners would need to earmark 1.6 percent of their capital per annum for the next decade for explicit dual-impact purposes: 1.6 percent of $220 trillion is $3.52 trillion. This pool of segregated assets would also need to be allocated appropriately, first and foremost with verified impact but also in a way such that our energy transition and broader economic and social needs are addressed comprehensively.

Now you understand why I put "solution" in quotations marks. The objective, mathematical truth is that the human family has all the money it needs to fashion the type of future we say we all want and need. But will we summon the collective will do it—and will we have the wisdom to allocate these funds appropriately?

A target of 1.6 percent per year is simultaneously ambitious and plausible. For example, a modest portion of assets now dedicated to indexed ESG and investment-grade bond strategies could be reallocated. As we've learned, these strategies and products now command more than $140 trillion in assets. A modest annual 2.5 percent reallocation from these pools alone would meet our estimated $3.5 trillion/1.6 percent goal. Figure 17.1 is a version of the same figure we saw in the last chapter, depicting various ESG equity investment strategies relative to their impacts. Rather than showing their return prospects relative to an index, however, the y-axis depicts the asset size of each strategy, of which there are three: indexed, active, and impact. This figure intuitively and convincingly shows that current ESG investment assets could be efficiently reallocated for more optimal, knowable impact. As we have repeatedly discussed, many indexed ESG strategies are valid risk–reward investment propositions. They should not be abandoned for the wrong purpose. Choosing the right index is among the most important decisions any investor must make. As we have also discussed, the best indexed ESG strategies will account for material ESG risks as systematically

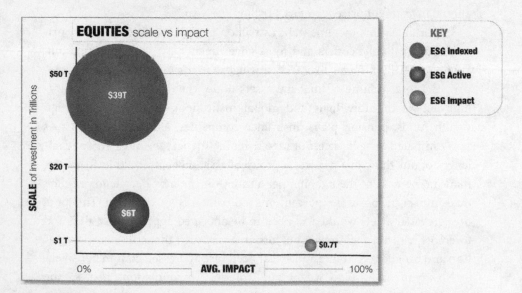

FIGURE 17.1. Current ESG equity allocations: size versus knowable impact.
Figure by the author.

as improving data sets allow. But double-bottom-line investing—or more colloquially, "doing well while doing good"—is best found in the ESG impact category. These strategies include both public and private equity strategies like the ones we've considered; for example, those of Bain Capital, Baillie Gifford, and TPG.

Similarly, there is a considerable opportunity to transition investment-grade fixed-income assets from their current category of low to zero impact to more impactful strategies, such as those now apparent in the rapidly growing green, social, and sustainable bond categories. As figure 17.2 shows, the investment-grade debt market is currently a $100 trillion market. A meaningful opportunity exists for sovereign and corporate issuers to dedicate significant future proceeds of their debt issuance to explicit SDG needs of both environmental and social natures. Institutional investors willing to rank verifiable impact higher in their investment objective functions could choose to support these efforts by reallocating portions of their current investment-grade holdings to ESG-blend or dedicated green, social, and sustainable bond strategies. In some ways, this reallocation strategy is even more compelling than those involving reallocations from public and private equity funds to ESG impact strategies. How so? For one, the investment-grade green, social, and sustainable bond markets have the advantage

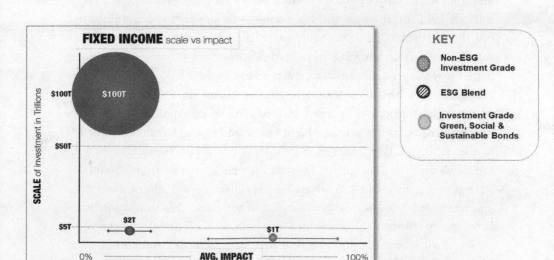

FIGURE 17.2. Investment-grade and impact fixed-income strategies by size.
Figure by the author.

of being far more targeted to specific, desired, and measurable outcomes. An investor could specifically choose whether and how their investments might lower emissions, improve income or gender inequality, or provide more low-income housing, to cite just three of many possible applications. Impact precision is possible. Investors could simultaneously choose their preferred jurisdictions for impact—from the south Bronx to South Africa or from Santiago, Chile, to Selma, Alabama. In other words, location precision is also possible. Green, social, and sustainable borrowing needs can be found in nearly every community. The debt markets are also now ideally positioned to help remediate specific community needs. Equally important, it very much appears global regulators and central bankers are committed to the proper functioning of global sustainable-debt markets over the long run. A significant component of the quantitative easing purchases by the European Central Bank over the past few years effectively supported the green bond market. Central bankers are also now actively considering what their role can and should be in muting the most destabilizing effects of carbon emissions and social unrest. Diligent, active support for green, social, and sustainable bond markets seems well within most central bank mandates and authority. Impact-minded fixed-income investors have the additional comfort of knowing the sanctity of these markets is already seen as a public good.

Similarly, the further deepening of the public impact equity market could and almost certainly would serve multiple social and environmental goals. Publicly held corporations and financial institutions that successfully carry out their sustainability missions in accordance with best business practices would benefit from knowing they could rely upon the full valuation of their firms as well as ready access to capital markets as growth opportunities present themselves. This includes micro-lending firms like Dvara Holdings or RDM, a global leader in recycled cartonboard which Apollo's impact team has nurtured. As companies like Penn Foster and Rodeo Dental are returned to the public markets, engaged shareholders could also provide them with the stewardship and incremental capital they need to grow from strength to strength. After all, public and private markets coexist in a single, unified financial ecosystem. The health and well-being of the *public* debt and equity markets directly influence the health and well-being of their *private* counterparts.

The most immediate and practical first step motivated institutional and high-net-worth investors could take in implementing my proposed 1.6 percent "solution" is to identify the verifiable level of impact that already resides within their existing portfolios. Following this crucial impact accounting exercise, they could then set incremental targets for the level of impact they wish to have in the future. For example, one impact objective an endowment or public pension plan may wish to have could involve linking their future investment plan to specific carbon reduction targets; another could be to target higher wages or explicit gender diversity goals in one's community. In all cases, though, *only impact that can be verifiably measured can be counted*. Admittedly, this process will be somewhat demanding for asset owners and their managers at the beginning. Still, the incremental effort needed would ultimately be rewarded verifiably with more inclusive, more sustainable economic growth. This shift in investment mindset may also appeal in a special way to those who have committed to net-zero portfolios. Shifting from a net-zero pledge to the more tangible 1.6 percent "solution" pledge could help solve two nettlesome problems at the same time.

As currently conceived, net-zero portfolios have no clear definition or broadly accepted accounting framework. Net-zero portfolios also need not involve any tangible social or environmental impact. After all, a portolio consisting of a small handful of net-zero companies is a net-zero portfolio, but to what end? The assumption that a net-zero portfolio helps the world get to a net-zero outcome is also merely that: an assumption. Without verified and measured impact, it's unclear what true environmental value a net-zero portfolio really contains. Note further that it is entirely possible for every endowment and pension plan in the world to have a net-zero portfolio without the globe being on a true net-zero path. How so? As we have seen, state-owned enterprises and privately held corporations will continue to supply all the oil and gas consumers demand no matter what happens to the ownership of publicly owned oil and gas companies. Stated more bluntly—Putin will do as he damn well pleases no matter how Cambridge decides to invest its endowment. Moreover, knowing how net-zero portfolios will ultimately be measured is also challenging and likely to remain so for years to come. If you are short Exxon or Chevron stock, should that be included as a carbon credit in your portfolio? If you buy a large tree farm or wind park from another owner, do you get to record a carbon credit even though no additional trees were planted or new wind mills put into use? When it comes to solving our social and environmental challenges, what matters most is net impact, not some simple carbon accounting scheme. Incremental investment *flows* matter, not *stock*. Why? Additional incremental flows are needed for real change and real impact: they alone connote true progress versus the past. This is why market-leading impact investing strategies require verifiable additionality. When one records an investment that claims to make a community or company more inclusive or sustainable, additionality data must be provided to back that claim up. The 1.6 percent "solution" pledge seeks to resolve these carbon and impact measurement conundrums simultaneously. As opposed to net-zero portfolios, the challenge of accounting for additionality is explictly demanded in the 1.6 percent "solution" pledge. All participating institutions and individuals must commit to reallocating unconstrained capital to products that include accredited, annual, incremental impact reports. As demand for impact products and strategies picks up, I assure you the private sector will become much more innovative and diligent in the accuracy of their impact accounting and reporting. It's what competitive markets do. In fact, crusty old Schumpeter himself will be actively at work, ensuring that more verifiable good is being generated while fees are kept down for discerning consumers. It's just that Shumpeterian creative *destruction* will be working in reverse. Perhaps we should call it Shumpeterian creative *construction*!

While not competely certain, a reallocation of 1.6 percent per year for the next decade could also be sufficiently gradual not to disrupt current market valuations or trigger unsustainable asset bubbles, though here considerable investment discernment would be needed. As can be seen in figure 17.3, recent impact investment allocations in the private debt and private equity markets now account for less than 10 percent of all allocations. In the public equity and public debt markets, impact allocations are even lower, at less than 1 percent of total capital committed. Investors who want their capital to do well and do good are not now at a loss for valuable and viable alternatives. As Casey Quirk predicted, and as European and other regulators are actively planning for, the impact investment markets are poised for further innovation, clarity, and growth. When asset owners start demanding more verifiable impact in their portfolios, asset managers will respond creatively and commensurately.

If institutional and high-net-worth investors were to reallocate 1.6 percent of their assets per year to genuine impact strategies spanning the debt, equity, infrastructure, and real-asset markets—public and private—global investors could do their part to keep carbon levels on track for net zero by 2050. They could also contribute tangibly to better health standards, greater sanitation and safe water sources, universal primary education, greater gender equity, heightened food security, more far-reaching telemedicine, and every other salutary objective envisioned by the UN's Sustainable Development Goals.

Impact Investing: The Opportunity

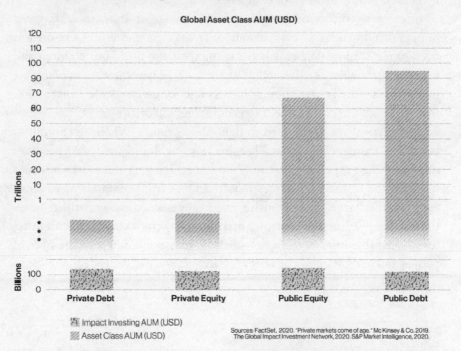

FIGURE 17.3. Impact investments now account for about 10 percent of the private debt and equity markets—and less than 1 percent of the public debt and equity markets. Significant additional allocations can be made to all impact asset strategies without exhausting the global impact opportunity set.
Data source: Deal Logic.

This seems to me the precise opportunity Robert Shiller anticipated when he wrote that investors are "among the most important stewards" and "vitally important players in the service of healthy and prosperous market democracies, all in service of the Good Society." More sophisticated and voluminous impact investing would undoubtedly accelerate our journey toward becoming the good society. We have all the money we need. We also have a clear understanding of the expanding opportunity set.

But perhaps increased impact investing is more than an opportunity. Unless and until investors assign more specific and measurable impact goals to their investments, it is highly unlikely that our ongoing emissions, environmental, gender equity, and economic justice objectives will receive the funding they require for resolution. Private investors are needed to fill the funding gaps public coffers cannot afford. Progress now depends more than ever upon private investment choices.

In other words, more than an *opportunity*, larger amounts of targeted impact investments may even be our collective *responsibility*. Asset owners need to invest more deliberately today to create the world our children and grandchildren deserve to inhabit. If we do not mobilize these types of impact investments with greater thoughtfulness and in greater scale, what might that say about our commitment to generations unborn?

Every parent wants to invest in their child's future. When we commit to being our brothers' and sisters' keepers, every child in the world becomes our own. The 1.6 percent "solution" or something very much like it would facilitate the financial resources required for a truly comprehensive strategy to turbocharge more inclusive, sustainable economic growth. A large, thoughtful, persistent reallocation of assets away from strategies with little-to-no verifiable impact, to those which demonstrably improve the lives of people and our environment is now entirely rational and needed.

CONCLUSION

(or How to Avert Our Failed Future)

I'm optimistic. I think we can pass from conquerors to stewards.

E. O. WILSON

This book endorses a vision of capitalism that works for people and planet. It also provides a viable action plan that would promote more inclusive, sustainable growth. There is an optimal path forward for the human family—one which could maximize human flourishing for centuries to come—but it isn't the easiest, nor the most likely. It also differs in many respects from the current path we are on. There are many practical things that we could do right now that would improve the lives of billions of people. Moreover, the risks of inaction are grave. . . .

In our failed future, decades of lagging economic growth and environmental neglect would relegate billions to miserable, hardscrabble living. Insufficient growth has always imperiled human tranquility. Given global populations are still expanding, inadequate growth in the years ahead would quickly cascade into declining living standards and rising social unrest. When economies fail, governments follow. Being broke is the most common trait of broken governments. On present course, few governments around the world will be able to take care of their aging populations. Comfortable retirements will remain possible in a handful of countries—Norway, Switzerland, perhaps the Netherlands. But in most other countries—including the United States, France, China, Brazil, India, and Japan—there is no viable plan now in place to support their aged citizens in just a few decades' time. These and many other countries are dangerously reliant upon demographically unsound "pay-as-you-go" pension funds; before they know it, their retirees will vastly outnumber their workers. How do countries survive when every worker needs to support two or three retirees, as well as themselves? Without greater forethought girded by much higher private savings, billions of the world's elderly not long from now will end up surviving on nothing more than meager monthly stipends. Hundreds of millions will be unable to afford adequate food, shelter, and health care. This crisis likely supersedes all others in terms of gravity, with the exception of heightened and prolonged military conflict.

But demographics aren't our only challenge. Our current course also invites multiple environmental disasters. Rather than reducing our carbon footprints, we are still allowing them to multiply carelessly. We are also over-mining the earth for essential minerals, like cobalt and lithium, simply to produce the batteries we now need. No energy source has a zero environmental footprint. On our present trajectory, it's entirely possible to see how 10–15 percent of the earth's landmass could become unlivable over time, victimized by unbearable temperatures, advancing deserts, rising seas, and/or strip-mining. Many densely-populated, seaside cities—from Miami to Jakarta, Dhaka to Lagos—could also become uninhabitable. Like hundreds of other coastal cities, their core infrastructure could well become too expensive to maintain due to devastating storms and recurrent floods. In certain extreme yet still conceivable scenarios, millions of desperate migrants may criss-cross the globe in search of better futures. How could it be otherwise? Risking one's life on treacherous maritime crossings and lethal barbed-wire fences is one's only rational choice when meaningful livelihoods are no longer possible at home.

When states fail, lawlessness reigns. When human beings refuse to regard others as their brothers and sisters, tribal identities reassert themselves. Stateless billionaires can secure their own armies and institute their own forms of law and order, divorcing themselves from the countries where they made their fortunes. Collective responsibility is a social good, yet one can easily imagine how living at sea ends up becoming the ultimate gated community for the ultra-rich.

There would be no globally respected reserve currency in our failed future, of course. Dollars, euros, renminbi, and yen would be regarded as empty paper promises. Even today's presumed financial saviors—cryptocurrencies—could well multiply in number before all becoming discredited. There are no real barriers to entry for cryptocurrencies. Rolling boom-bust cycles, unenforceable transnational regulation, and recurrent fraud will likely prohibit cryptocurrencies from commanding sufficiently broad, sustained consumer confidence. It is hard to imagine how any official currency could be trusted in our worst-case, failed future, digital or otherwise. Instead, food, fresh water, firearms large and small, and essential commodities like oil, gas, cobalt, copper, and nickel may become the most commonly used mediums of exchange. Bartering would effectively displace blockchain.

In our failed future, Canada, Mongolia, Finland, Sweden, Russia, North Korea, Tajikistan, Kyrgystan, the Czech Republic, Argentina, Alaska and Iceland—i.e., all colder clime states—would gain *relative* competitive strengths versus their less well-situated counterparts in higher temperature zones. This said, they, too, would struggle with extreme weather events, unreliable food and water supplies, and unsafe borders. Constant threats of nuclear, chemical, and biological warfare could easily make them and many other nation states treacherous places to live. Such is a lawless world. The broadscale risks of neglect and inaction would be grave for everyone. As miseries mount, there would be nowhere to hide. All the obvious opportunities for more inclusive, sustainable growth that we painstakingly envisioned and explored above may be ignominiously ignored. Rather than a "spring of hope," in our failed future, humanity would face a never-ending "winter of despair."

But like famed biologist E. O. Wilson, I am more optimistic than this. Much more. I don't believe the extreme scenarios I just described will come to pass. Not at all. I'm optimistic because I believe a large majority of engaged and enlightened citizens will insist upon evolving from conquering our environment and each other to become more faithful stewards and mindful communitarians. I am optimistic because hope and progress wither in its absence, and no meaningful life can be lived without hope. And I'm optimistic because I am a *pragmatist*. Study history and you'll see optimists have always proven pessimists wrong. It is always darkest before the dawn.

In our best future, private markets still drive asset allocation and investment; they just do so more successfully because they have been mindfully guided. It's not just that antitrust and competition laws will continue to be upheld; a collective commitment to our children and grandchildren will simply prove too strong for wanton negligence and corruption to reign. Many current and soon-to-emerge innovations will also have revolutionary, positive impacts, once again rescuing the human family from what we now wrongly perceive to be existential threats. Such innovations have always proven Malthusian doomsayers wrong. Devastating and costly acts of nature caused by climate change will almost certainly become more common—to be frank, at this stage, they have become impossible to avoid. Still, how we adapt to and help mitigate these recurrent, extreme natural events can and almost certainly will exceed current projections. Among other things, our infrastructure will better anticipate climate change and be built to last. In other words, *we will defeat climate change in large part by anticipating it.* As Singapore is now demonstrating, higher seas brought by climate change need not be an extinction threat for low-lying countries. Intelligent human beings have always found ways to overcome our changing, natural environment. This said, higher seas are a threat for far too many in low-income countries. One of the greatest challenges for the just transition involves better preparations, warnings, and protections for those who are most vulnerable—something existing communications technologies are already doing.

At the forefront of the transformational innovations I see in our enlightened future are thousands of technological breakthroughs. Not only will we electrify transportation and decarbonize the grid faster than projected, but a vast majority of companies, communities, and countries will embrace *credible* plans to be part of the circular economy. Among these surprises will be new forms of carbon capture and storage, as well as growing numbers of personal, corporate, and national commitments to operate more sustainably. Among other successes, these same institutions and individuals will help plant hundreds of billions of trees, an initiative that alone could sequester two hundred gigatons of carbon over their lifetimes. In addition to cleaning up industry, revolutionary technology breakthroughs will reduce the environmental damage modern food production perpetuates, with regenerative rather than depletive techniques going mainstream. Driving all these efforts for greater sustainability will be legions of mindful consumers who consistently align their wallets with their children's best interests; after all, consumers ultimately drive business decisions. More mindful consumption will herald new, sustainable growth industries and products. In our enlightened future, personal commitments to get to net zero will infuse many elements of our lives, just as self-sacrifice and ubiquitous "no-lose" mentalities drove the Greatest Generation to victory in World War II and helped the massively out-gunned Ukrainian army inspire the world.

As we benefit from the efforts of a collection of farsighted scientists, global investors, energy industry experts, and policy-makers, beginning in 2022, I also believe our current, counterproductive obsession with oil and gas *producers* will transform into a far more fruitful and honest discussion about how to power our future growth reliably and sustainably. After all, if we can achieve net-zero emissions, talk of an "energy transition" becomes moot. Importantly, more nuclear power will be brought into the mix. Oil and gas will come to be seen as they are—i.e., necessary fuel sources that must power many companies, countries, and broader human progress for centuries to come. The last few years have revealed painfully how reckless obsessions with only one of our greatest challenges—excessive carbon emissions—cannot and will not be solved by perpetually shaming the providers of the energy we need and upon which we will continue to rely. Our collective focus must prioritize reliability, security, and affordability no less than environmental sustainability. It is not in humanity's best interest to abandon our existing mix of energy sources when other, cleaner, and more cost-effective solutions are not yet dependably available. It is equally senseless to depend on an autocratic Russia for anything, let alone our economic well-being. Putin's oil and dynastic ambitions are far too dirty in every sense. The human family needs abundant, reliable, affordable, safe, *and* clean energy systems. Moreover, these systems are needed at both individual community and national levels. We should not and must not expect any nation-state to leave themselves vulnerable to the whims of weather or the beneficence of others as they create the energy system they need for their own people to thrive and prosper. Energy security must work locally and nationally before it has any chance of working globally.

For this reason, in our enlightened future, a group of institutional investors led by sovereign wealth funds, public pension plans, and multilateral institutions like the World Bank and other development banks will come together on a comprehensive solution to help emerging market countries transition from dirty coal and other heavily carbon-based power plants to cleaner, more reliable, and more sustainable energy sources. They will do so by finding more innovative ways to share sovereign credit and currency risks, unleashing trillions of dollars of much needed public and private capital that ultimately turn China, India, Indonesia, Vietnam, Poland, Turkey, and other heavily carbon-dependent countries into cleaner, net zero economies years ahead of current projections. Their combined farsighted efforts will not only foster greener outcomes but also advance more inclusive economic growth for all of their citizens. Energy insecurity is a direct cause of poverty. When inclusive, sustainable growth is explicitly accepted as the tripartite objective of every community and country, more affordable, reliable, and cleaner energy becomes axiomatic.

Concomitant with mounting clean energy successes, in our enlightened future, a group of public- and private-market investment firms like Bain

Capital, TPG, Apollo, LeapFrog, Fidelity, BlackRock, Amundi, Baillie Gifford, and others will find replicable and scalable success in generating significant shareholder value through the identification and reward of proven circular economy applications. Thousands of corporations seeking savings through more sustainable business practices will be further aided by organizations like the Environmental Defense Fund and the Shared Value Initiative. By applying proven, cost-saving techniques, these leading investment houses and advisory firms will not only make senseless corporate waste a forgotten relic of centuries past, but by significantly lowering operating costs and improving profits, they will also help generate better returns for their clients, resulting in more fully funded pensions. Investing in sustainability will help hundreds of millions of pensioners avert want and achieve real comfort. Along with Blackstone, Starwood, Oaktree, Brookfield, and other leading real asset investment firms, global commercial real estate, infrastructure, and logistics operations will also continue to undergo significant operational, lower-carbon, cost-saving revolutions. Sustainable environmental footprints will become synonymous with value creation, in other words. For this reason, tens of thousands of high-quality, affordable retirement villages will be built in environmentally sound, less densely populated areas, creating safe, zero-carbon communities where hundreds of millions of septuagenarians and octogenarians will live out their last years with comfort and meaningful, human connection. Investors in these projects will also enjoy market returns well in excess of inflation, enabling their capital to increase in value while creating much needed social infrastructure—the ultimate model of doing well while doing good.

Private enterprise will also become a larger contributor to social progress. The goodwill and commitments of the Business Roundtable to foster more inclusive, sustainable growth will continue to multiply in our best future. Indeed, more companies than ever will commit to doing more good. Some will follow JP Morgan's new Policy Center, using evidence-based policy solutions in their unique areas of expertise to advance opportunity. After all, JP Morgan's *Second Chance Agenda*—which helps Americans who have an arrest or conviction record reenter the work force—is one of hundreds of examples of how more mindful corporations can do well and do good. By partnering with exemplar of hope programs—like Operation HOPE's One Million Black Businesses (1MBB)—many banks and other financial intermediaries will also find new ways to better their communities and benefit their shareholders at the same time. Given all these programs are "hand-ups" rather than "hand-outs," they will improve the lives of their givers no less than their recipients.

Of course, public policy will not stand still in our emerging, inspirational future of accelerating inclusive, sustainable growth. Citizens will demand and ultimately be rewarded with strong, visionary leadership that excels at straight talk and GSD (getting *stuff* done!). Visionary leaders in our better future will be battle-ready crisis managers. They will also understand that, when society wants less of something, you tax it; when society wants more of something, you leave it alone or incentivize it. For this reason, in our enlightened future, a universally applied carbon tax—small at first though gradually-rising—will become law, with border offsets for the products and services of countries like Russia or China who fail to tax the carbon-content of their goods commensurately, or which otherwise flout broadly agreed behavioral standards. If we want less carbon over time, we must tax it accordingly; if we do not wish to enrich expansionist, autocratic regimes, we must stop doing business with them. Taxes on excessive consumption and luxury items will also be in place, ensuring those who can pay more do. Importantly, confiscatory income taxes will be broadly *avoided* in our successful future, with no one incurring a marginal rate on regular income in excess of 50 percent. Taxation on all short-term financial speculation will rise sharply, while taxes on long-term investment and capital formation will be gradually lowered over longer periods. In fact, all capital gains taxes will be phased out to zero after five years. Investing for better long-term outcomes is crucial. In our successful future, it will finally be incentivized accordingly.

Similarly, a modest, yet unavoidable wealth tax will become law for all estates above an inflation-indexed limit in our successful future, displacing ineffective and easily avoided estate taxes. At less than one percent per year on all assets, this modest tax will end up being too small for anyone to actively contravene, yet highly efficacious for generating significant revenue from large family fortunes when assessed over decades. Some versions of universal basic income, universal pre-kindergarten through twelfth-grade education, and basic universal health protections will also be put in place in every middle- and higher-income country. Support programs like these are essential to give every child a fair start in life, as John Rawls rightly envisioned. This said, income support programs will never become so comfortable nor so generous as to disincentivize their recipients from wanting more. Incentives matter. Aligning them properly is the most important economic variable driving communal success over failure. Indeed, proper incentives are why industriousness, innovation, long-term investment, personal thrift, and bountiful charitable activities will be widespread in our most enlightened future. Unless and until our social service programs align with human nature as it really is—that is, for meaning-seeking creatures who derive their greatest fulfillment from utilizing their God-given gifts rather than becoming dependent upon perpetual, unfunded and unearned benefits—human flourishing can never be maximized. Indeed, if we do not retain individual responsibility and common virtues as the cornerstones of our communities, humans won't be able to have lives of consequence and value, let alone material abundance. Industriousness is the cornerstone of all prosperity. Leisure is only possible through prosperity's perpetuation. That said, if we do not simultaneously help those who cannot help themselves, our sacred responsibility to be our brothers' and sisters' keepers will have been abrogated, and our communities will be far less rich and our lives far less rewarding. The fastest and most certain way to achieve personal fulfillment is by helping others achieve theirs. The richest and happiest communities are those where every member flourishes, most especially their least fortuned.

Most pertinently in the context of this book, in our enlightened future, capital for urgent environmental, social, and economic needs will not be in short supply. This is because every institutional and high-net-worth individual will reflect upon the existing level of impact in their portfolios and their ability to allocate some portion of their assets to more impactful investment outcomes. Special attention will be given to the issues that matter most in one's *local* communities; after all, our problems can only be resolved community-by-community. As growing numbers of asset owners adopt their own optimal version of the 1.6 percent "solution," green, social, and sustainability bonds will become much larger allocations in many fixed income portfolios, most especially insurance companies and pension plans. As actuarial-driven investors, these institutions well understand credit risk and the need for more inclusive, sustainable outcomes. As with public pension plans, it will prove natural for insurance companies to lean more into income-producing, purpose-driven investment opportunities. Central banks will also make more meaningful allocations to G20 sovereign green and sustainable bonds with their foreign exchange reserve assets. All investors will have greater impact.

Concomitantly, in our best future, many university endowments will augment their widespread, and likely ineffective, net-zero investment commitments by adding more explicit impact investment objectives. Working with organizations like the Global Impact Investment Network, the Ellen

MacArthur Foundation, the Global Steering Group for Impact Investment, the Impact Management Project, and Climate Action 100+, along with index giants like MSCI, Bloomberg, and S&P, this enlightened group of leading university endowments will also help establish best-practice standards for impact *measurement*. As they do, they will be joined by an astonishingly high percentage of family offices who are ready to leapfrog the 1.6 percent "solution" movement, allocating as much as 20 percent of their wealth to impact, or even more. Because of their strong, individual mission-driven beliefs, multiple family offices and high net worth individuals will give amplified attention to forgotten community needs and other, urgent social goals, including recidivism, literacy rates, prison reform, economic mobility, and maternal health care. And in addition to their partnerships with development banks to decarbonize energy grids in many emerging market countries, a uniquely enlightened group of sovereign wealth funds will pledge their considerable influence as well as meaningful portions of their assets to educate future generations of globally minded citizens from 195 countries (including the *State* of Palestine and the Holy See). Their goal? To optimize transnational cooperation wherever it is most needed. More than any other investors, sovereign wealth funds understand the ways in which nations must ultimately come together as a single family. They understand cross-border issues like pandemics, cleaner oceans, breathable air, immigration, and global trade can only be addressed collectively.

Complementing all of these scaled-out impact investment innovations and extraordinary, public policy breakthroughs will be a side-by-side, positively reinforcing revolution in private philanthropy. As growing numbers of exemplars of hope are identified, their proven methodologies will be scaled. Darren Walker of the Ford Foundation was visionary when he used his institution's considerable financial assets in a crisis to accelerate urgently needed relief. Why wait to address future problems when resources are readily available and so much needs to be done right now? Combined with the rigorous insights of Notre Dame's Wilson Sheehan Lab for Economic Opportunity, Columbia's Tamer Center for Social Enterprise, and the MIT-affiliated Abdul Latif Jameel Poverty Action Lab (J-PAL), many scientifically sound solutions for reducing poverty and improving economic mobility will also be identified and refined for mass adoption. Legions of philanthropic institutions with combined assets of more than $2 trillion will bond together and sign their own version of the "billionaires' pledge" (perhaps to be called the "trillionaires' pledge"). They will promise to use their considerable resources to implement scalable solutions in the immediate months and years to come, rather than decades away. Just as Mary Barra inspirationally committed GM to "zero crashes, zero emissions, and zero congestion," these visionary and philosophically aligned philanthropies will promise zero waste, zero want, zero injustice, and *zero unloved children.*

And in a historic act of multilateralism, the likes of which humanity has never known in peacetime, the G20 will miraculously coalesce. Through successive, collaborative presidencies that include the United States, China, India, France, Mexico, South Africa, Japan, Indonesia, Canada, the United Kingdom, Saudi Arabia, Germany, Australia, Brazil, and a Putin-free Russia, the G20 will collegially map out a future in which the human family can genuinely thrive as one. Among other breakthroughs, these nations will reimagine and modernize our current, anachronistic collection of Bretton Woods and UN institutions, modernizing them for the century ahead rather than leaving them as increasingly ineffective relics of the century behind. The goals of these reforms will include zero tolerance for corruption, military despots, money laundering, and the proliferation of all weapons of mass destruction; adequate shared pools of capital for sovereign debt restructuring, multilateral concessionary lending and global health needs (including funding to help poor countries fight future pandemics), as well as mutually agreed rules governing how these resources are dispersed; revitalized, inviolable respect for national borders and the rights of all people to live under the political systems they choose for themselves; recognized and empowered global courts for the fair adjudication of trade disputes as well as the prosecution of global criminals; and a new, global reserve currency standard that meets the necessary requirements of intergenerational confidence, efficiency, fairness and enforceability.

In our enlightened future, optimists will triumph, human ingenuity will emerge victorious, and eternal virtues and principles will have been reaffirmed and elevated. Evil will not have been irradicated; as we all know, it never can be. But the principles for promoting Aristotelian fairness—that is, fostering social equanimity by correcting all that is inequitable—will have been more broadly embraced and codified. As such, the commitment to fight all forms of injustice will grow stronger.

"Overly optimistic?" you may now be asking yourself. Well, yes, perhaps a bit.

But it would certainly beat our failed future—and for that reason alone, I think we should give it a try.

ACKNOWLEDGMENTS

When I began this book in the spring of 2021, I did not realize how much its completion would depend upon the kindness and superior insights of so many others. Over the course of its drafting and redrafting, I repeatedly called upon the generosity and greater wisdom of family members, colleagues, friends, and the team at Columbia University Press. Space permits me only to thank the most important.

While I have long benefitted from the expertise of dozens of BlackRock colleagues, for the purposes of this publication, I am especially grateful to Professor Andrew Ang, Jim Badenhausen, Rachael Banton, Scott Barber, Brian Beades, Susan Beadle, Dalia Blass, Paul Bodnar, Sandy Boss, Caroline Crozat, Annelise Eschmann, Thomas Fekete, Larry Fink, Lili Forouraghi, Laura Hildner, Mark Hume, Rich Kushel, Nitin Malik, Mark McCombe, Chris Meade, Matt Mosca, Salim Ramji, Jack Reerink, Eric Rice, Rick Rieder, Ned Rosenman, Ashley Schulten, Laura Segafredo, Mark Spizner, Laura Figlina Stahl, Quyen Tran, Peter Vaughan, Dan Waltcher, and Max Zamor. My friend and confidante Sarah Hallac merits special mention for her unflagging personal and professional support. I have learned many valuable life lessons from Sarah and her late husband Charlie, who served as BlackRock's co-president. To be completely clear, I believe BlackRock has done more good for more people on the planet, as well as for its future generations, than any other financial services firm in history. I also have hope it will continue to do so. While special credit for BlackRock's unrivaled financial achievements and socioeconomic contributions belongs to its founders and Group Executive Committee led by Larry Fink, Larry and all his senior partners know even more credit extends to BlackRock's 18,000 current employees as well as a growing alumni network, of whom I am now one. This said, neither BlackRock nor the individuals named above should be considered responsible for the contents of this book. Its opinions, conclusions, errors, and omissions are mine alone.

Other colleagues and friends whose counsel proved valuable in producing this work include Professor Jennifer Aaker, Fr. George Augustin, Marc Becker, Brian Bedell, Afsaneh Beschloss, Kevin Boucher, Matthew Breitfelder, Archbishop Charles Brown, John Hope Bryant, Oren Cass, Joseph Chi, Joe Connelly, Todd Cook, Senator Joe Donnelly, Stuart Dunbar, Charles Ewald, Maggie Fergusson, Professor Stanley Fischer, Professor Mary Ann Glendon, Senator Phil Gramm, Wendy Gramm, Jon Gray, Britt Harris, Rikki Jump, Karen Karniol-Tambour, Cardinal Walter Kaspar, Tara Kenney, George Magnus, Scott Malpass, Paul McCulley, Blake Murphy, Geoffrey Okamoto, Mike Pyle, Professor Emma Rasiel, Steve Reifenberg, Matthew Rutherford, Speaker Paul Ryan, Erik Schatzker, Deborah Spar, Savita Subramanian, Andrew Sullivan, Professor Jim Sullivan, Scott Tinker, Paul Tudor Jones, Jeffrey Ubben, Krae Van Sickle, Susan Webster, Fr. Oliver Williams CSC, and Spencer Zwick. I must also highlight the profound influence the late, legendary president of the University of Notre Dame, Fr. Theodore M. Hesburgh, CSC, has had on my Irish stubbornness and commitment to fashion a better world. We all have much to learn from Fr. Ted, beginning with his devout faith and his intolerance for injustice.

There would be no book without the constant guidance of Myles Thompson of Columbia University Press, or the work of his indefatigable colleague Brian Smith. Sheri Gilbert diligently chased down dozens of reprint permissions. Meredith Howard and Robin Massey helped me plan for its launch. Kalie Koscielak helped rid it of many errors.

As one of eight children, all of my brothers and sisters generously offered their opinions and suggestions, with my youngest brother Mark even offering to appear as a centerfold (check back for a second edition!). My brothers Tim and Larry, and my sister Carol read and re-read drafts, providing helpful suggestions that benefited the final work tremendously. Many of my nieces and nephews also read drafts and provided comments. This said, my eldest niece—Maren Keeley, whose professional life has been largely dedicated to more conscious capitalism—provided detailed, page-by-page improvements. Her energies and visions inspired me to be more focused and bold in my conclusions. Our shared passions are why we are now committed to working together, to make all these words matter.

Last but proverbially not least, my life partner Saskia Bory-Keeley and our two sons Julian and Calum put up with hundreds of hours of my dominating cocktail conversations, being late to the dinner table and up VERY early, typing away in my home office. My gratitude for their presence in my life cannot be put into words—as I hope this book's dedication makes plain.

EXEMPLARS OF HOPE

An exemplar of hope is an NGO or other civic social organization that has a scalable program or methodology that verifiably promotes inclusive, sustainable growth. To qualify as an exemplar of hope, the methodologies of the NGO or civic social organization cannot promote inclusivity or sustainability at the expense of growth or promote inclusivity or growth in an unsustainable manner. All three goals—inclusivity, sustainability, and economic growth—must be advanced by one or more of its core programs.

The list that follows is illustrative and, I must confess, overly U.S.-centric. For a current, more comprehensive list, please visit www.1pointsix.com. If you would like to nominate an organization to be an exemplar of hope, please send the name of the organization, a description of its methodology, and its website address to terry@1pointsix.com. Successful nominations will be added following a vetting procedure and permissions process.

Alamo Academies familiarizes high school students with STEM careers with excellent employment prospects. Program graduates have an average starting salary of $42,000 (https://www.alamoacademies.com).

Beyond Differences promotes inclusion through social-emotional learning (SEL), a transformative proprietary curriculum that has been used in more than 8,500 schools across all fifty U.S. states. SEL has helped more than 3.5 million students move beyond isolation and loneliness toward community (https://www.beyond differences.org).

Botvin LifeSkills Training is a three-year universal substance abuse and violence protection program targeted at middle, junior, and high school students. LST has been proven to cut teen drug use by 75 percent, tobacco use by 87 percent, and alcohol use by 60 percent (https://www.lifeskillstraining.com).

C40 Cities is a group of ninety-seven cities around the world dedicated to sharing best practices for reducing greenhouse gas emissions and mitigating climate risks (https://www.C40.org).

Career Academies are accredited counseling programs applied in small learning communities at the high school level across the United States. Their "school-within-school" programs provide actionable advice on local career opportunities as well as the technical skills needed to transition to the workplace upon graduation (https://www.evidencedbasedprograms.org/programs/career-academies).

Catholic Charities Fort Worth provides service to those in need, advancing compassion and justice in the structures of society while encouraging all people of good will to do the same. Their Out of Poverty Pathways program is strictly outcome driven and evidence based. This program has helped tens of thousands of families across CCFW's twenty-eight-county diocese (https://www.catholiccharities fortworth.org).

Charter for Compassion was founded by the religious scholar Karen Armstrong. The CFC invites every human being to co-create a world that is peaceful, kind, inclusive, and rewarding. Charter signatories believe compassion can and should infuse every human relationship and approach to societal life (https://www.charter forcompassion.org).

Critical Time Intervention provides short-term interventions for individuals who have been living in shelters and homeless centers, providing them the help they need to return to normal life. Longer-term goals include reintegration and independence of community support networks (https://www.criticaltime.org).

EMPath (Economic Mobility Pathways) provides intensive counseling to help people living in poverty and those at risk of living in poverty achieve economic independence. EMPath also provides affiliated institutions and organizations with the tools they need to implement their proven programs (https://www.empathways .org).

The Environmental Defense Fund ranks among the world's leading environmental organizations, with more than 2.5 million members and a proven fifty-year record of making the environment safer and healthier. They advise companies, communities, and countries on sustainability best practices (http://www.edf.org).

FAME USA (the Federation for Advanced Manufacturing Education) provides work-force development skills, helping to place 85 percent of its participants with more than four hundred companies across twelve U.S. states (https://www.fame-usa.com).

The George Kaiser Family Foundation is dedicated to providing every child in and around Tulsa, Oklahoma, an equal opportunity to succeed in life. They employ a wide range of local and national public–private partnerships to scale evidenced-based practices that narrow opportunity gaps to age eight (https://www.gkff.org).

The Good Things Foundation is a UK-based social change charity that improves people's lives by helping to close the digital divide. Their Make It Click program provides free courses, tools, and templates across the United Kingdom and around the globe to improve digital literacy (https://www.goodthingsfoundation.org).

The Green Belt Movement was founded in 1977 by Professor Wangari Maathai. GBM has planted tens of millions of trees across Africa, working at local, national, and international levels to promote environmental conservation, climate resilience, and gender equality (https://www.greenbeltmovement.org).

The International Federation of Red Cross and Red Crescent Societies hopes to inspire, encourage, facilitate, and promote humanitarian activities through 192 national societies located in countries all around the world. They do so with a view to preventing and alleviating human suffering and contributing to the main-tenance and promotion of human dignity and peace (https://www.ifrc.org).

KIPP Public Schools are a network of hundreds of tuition-free public charter schools across the United States, with more than 10,000 educators and 160,000 students and alumni. KIPP affiliation requires specific governance and pedagogi-cal tools. Because of excess demand for their services, admission is often by lot-tery (https://www.kipp.org).

Lutheran Services in America runs one of the largest health and human services networks in the United States. They help care for more than six million Ameri-cans in 1,400 communities across the country, as well as a large number of immi-grants and refugees. Lutheran Services has more than 150,000 active volunteers (https://www.lutheranservices.org).

Nurse-Family Partnership is a nationwide community health program that empow-ers first-time moms to improve their lives and the lives of their babies through the dedicated use of best practices. They envision a future in which all children are healthy, families thrive, and the cycle of poverty is broken (https://www.nurse familypartnership.org).

Operation HOPE is dedicated to expanding economic opportunities by making free enterprise work better for everyone. Their primary focus is on promoting financial dignity, literacy, and inclusion among the underserved. Their Financial

Literacy for All movement is developing content for universal use in America and around the world (https://www.operationhope.org).

Ready, Willing & Able (a.k.a. the Doe Fund) provides men with a history of incarceration, homelessness, and unemployment with paid work and career development services, providing proven pathways from insecurity and want to a full-time job and a permanent home. Their Work Works model verifiably reduces recidivism and has the potential to save hundreds of millions of public program dollars while improving tens of thousands of lives (https://www.doe.org).

Robin Hood has successfully fought poverty in New York City and its environs for more than thirty years. During the COVID-19 pandemic, they prioritized getting families back on their feet, kids back on track, and able-bodied New Yorkers back to work (https://www.robinhood.org).

Rumie was founded by the former BlackRock executive Tariq Fancy. Rumie is a Canadian NGO and technology company that has provided educational assets and tools to more than three hundred thousand active learners in more than 191 countries. Rumie has curated more than 1,100 free and open courses and hopes to reach twenty million learners by 2023 (https://www.rumie.org).

Year Up is dedicated to ensuring every young adult reaches their full potential, closing the opportunity divide by providing access to skill attainment and proven personal empowerment programs (https://www.yearup.org).

YMCA is a worldwide youth organization headquartered in Geneva, Switzerland, that serves more than one hundred million beneficiaries in 120 countries. It strives to put Christian principles like compassion, patience, self-control, and kindness into practice while developing healthy bodies, minds, and spirits (https://www.ymca.int).

NOTES

INTRODUCTION

1. As I wrote in the ruins of the global financial crisis, "As a man of conscience, I have been greatly pained by the economic and financial headlines of the past eighteen months. I have wondered whether my life has been inadvertently given over to ignoble pursuit, if not outright evil. Are Notre Dame's sons and daughters on Wall Street, myself among them, somehow culpable for the mess our country finds itself in? If so, what in heaven's name should we do about it?" Terrence Keeley, "Eye of the Needle," *Notre Dame Magazine*, Summer 2009. This article won a General Interest award from the Catholic Press Association in 2010.
2. Robert J. Shiller, *Finance and the Good Society* (Princeton, NJ: Princeton University Press, 2012), 27.

1. THE STAKES

1. Brad Cornell, "The ESG Concept Has Been Overhyped and Oversold," *Financial Times*, July 15, 2020.
2. Tariq Fancy, "Financial World Greenwashing the Public with Deadly Distraction in Sustainable Investing Practices," *USA Today*, March 16, 2021.
3. Let me make clear why I disagree with Tariq Fancy's broadscale "duping" charge. Responsibility for capital allocation lies first and foremost with the owners of capital. There is no evidence to suggest thousands of investors have somehow been blindly misled. Moreover, asset managers are not culpable for the informed decisions asset owners make. Investors hire asset managers to achieve specific goals. If those goals are not being met, asset owners have the right to fire their managers. If certain ESG products are not generating the results owners need and want, those investors have the

opportunity to course correct. And if the whole system is not generating the results society wants, systemic change is needed.

4. As Professor Mazzucato persuasively writes, "Only government has the capacity to bring about transformation on the scale needed." That government "must first be equipped with the tools, organization and culture it needs to drive a mission-oriented approach." In other words, we must first create enlightened leaders and then give them the requisite powers to effect change. It's a compelling idea, but one wonders how it is any different from what China is already doing. Is Sino-capitalism verifiably superior to Western capitalism? I discuss this question in the next chapter. See Mariana Mazzucato, *Mission Economy: A Moonshot Guide to Changing Capitalism* (New York: HarperCollins, 2021).

5. Inexplicably, the damning role of the former House Financial Services chairman, the Democrat Barney Frank, in planting the seeds and then fanning the flames of the subprime crisis has never been properly told. "I want to roll the dice a little bit more in this situation towards subsidized housing," Barney famously remarked in 2003, effectively endorsing hundreds of billions' worth of additional subprime loans that ended up on Fannie's and Freddie's balance sheets. Those dice came up snake eyes only a few years later. As a general rule, governments should not allocate credit. When politics influence actuarial and credit analysis, less sound financial decisions follow.

6. Peter H. Lindert and Jeffrey G. Williamson, *American Incomes 1774–1860*, NBER Working Paper 18396 (Cambridge, MA: National Bureau of Economic Research, September 2012).

7. The Brookings Institution reports that the number of individuals living in extreme poverty fell from 1.9 billion in 1990 to 648 million in 2019. They had projected it would fall further to 537 million in the next few years, but COVID-19 has set back this trend. Not long before Russia's senseless invasion of the Ukraine, they projected that global poverty will not fall below 2019 levels before 2023 and that 588 million will be living in extreme poverty in 2030. Russia now imperils every human being, rich and poor—including Russian citizens.

8. Global population growth is itself an unsettled debate. The respected British medical journal *The Lancet* recently projected a peak global population of 9.73 billion by 2064, up from about 7.8 billion in 2021. This is somewhat lower and earlier than prior UN estimates, which put peak global population at nearly 11 billion by around 2100. The trajectory is nevertheless clear: higher. At a minimum, we should expect and plan for an approximately 25 percent increase in members of the human family this century. Each will require food, shelter, education, and meaningful work.

9. For a more compelling, system-wide analysis of the positive, bottom-line impact of worker satisfaction, see Clement Bellet, Jan-Emmanuel De Neve, and George Ward, "Does Employee Happiness Have an Impact on Productivity?" Saïd Business School WP 2019–13, *SSRN*, October 14, 2019, http://dx.doi.org/10.2139/ssrn.3470734.

10. In his new book *Unsettled*, the noted physicist Steven Koonin pushes back on some of these claims, arguing "the net economic impact of human-induced climate change will be minimal at least through the end of this century." Dr. Koonin does not deny humans are changing the composition of the atmosphere at a record pace or that the next century could ultimately be much warmer. Rather, he believes "socio-technical obstacles to reducing CO_2 emissions make it likely that human influences on the climate will not be stabilized let alone reduced this century." Instead of spending trillions to alter the nature and amount of energy we use, Koonin advocates policies that would help societies adapt to the negative effluents of our current energy sources. The debate about spending our

limited financial resources on *mitigation* versus *adaptation* will be among the most consequential in the years to come. See Steven E. Koonin, *Unsettled: What Climate Science Tells Us, What It Doesn't, and Why It Matters* (Dallas, TX: BenBella, 2021).

11. For a sobering read on what can happen to societies when income levels become excessively disparate, see Walter Scheidel, *The Great Leveler: Violence and the History of Inequality from the Stone Age to the Twenty-First Century* (Princeton, NJ: Princeton University Press, 2018).

12. My dear friend and longtime mentor, Senator Phil Gramm of Texas, can data you to tears on this topic. Start with Phil Gramm, "The Truth About Income Inequality," *Wall Street Journal*, November 3, 2019. Then Phil Gramm, "Incredible Shrinking Income Inequality," *Wall Street Journal*, March 24, 2021. Warning: if you choose to debate Senator Gramm on this topic, you will lose. He's armed with facts, seldom wrong—and simply unrelenting.

13. John Rawls, *A Theory of Justice* (Cambridge, MA: Harvard University Press, 1971).

14. Katharine Graham became the first female CEO of a Fortune 500 company in 1972, after her husband was appointed CEO by her father but then committed suicide. Her tenure at the helm of the *Washington Post* was defined in no small part by her handling of the Watergate scandal. Nixon's attorney general, John Mitchell, said of the *Washington Post*'s Watergate coverage that "Katie Graham's gonna get her tit caught in a big fat wringer if it gets published." But it was AG Mitchell's and President Nixon's tits that got wrung in the end. Katharine Graham not only survived; she and the *Washington Post* thrived.

15. To be clear, there is no reason to presume continued human progress of a linear or even quasilinear nature. There have been many multidecade setbacks in modern societies, such as those in New York City from the mid-1960s to the mid-1990s and in Germany during the first half of the twentieth century. Whole civilizations have also moved from high levels of development to complete chaos in relatively short periods, only to stay there. Famously, ancient Rome achieved unprecedented economic growth under a succession of five emperors over two hundred years—but devolved into chaotic ruin in the two centuries that followed the death of Marcus Aurelius in 180 CE. No Western society managed to return to Rome's economic and political level until systematic advances in European agriculture and the Enlightenment occurred, some 1,400 years later. If we make serious mistakes now, the entire human family could easily spend lost decades or even centuries in despair.

2. STAKEHOLDERS VERSUS SHAREHOLDERS

1. Justice Russell Ostrander for the majority, Dodge v. Ford Motor Company, 204 Mich. 459,170 N.W. 668 (February 1919).

2. To be clear, Henry Ford was not a man of limitless virtue. His anti-Semitic sentiments were well known and all too often on display, including in his public support for Adolf Hitler until World War II began.

3. Adolf A. Berle Jr., "For Whom Corporate Managers Are Trustees: A Note," *Harvard Law Review* 45, no. 8 (June 1932): 1365, 1367, 1372.

4. This chapter's opening quote comes from "Stakeholder Capitalism vs. Milton Friedman: A Discussion with Darren Walker and James Stewart." This conversation took place at the Aspen Institute in Washington, DC, on September 16, 2020. Darren Walker is one of the more eloquent and ardent critics of shareholder capitalism. The foundation

Mr. Walker oversees was established in 1936 with an initial gift of $25,000 from Henry Ford's son, Edsel. The family gave the foundation an additional gift of $250,000 of Ford Motor stock in the mid-1940s, making the Ford Foundation then the largest philanthropic foundation in the world. Its stated mission is to reduce poverty and injustice, strengthen democratic values, promote international cooperation, and advance human achievement—that is, to do all those things stakeholder capitalists might cherish most. But for Henry Ford's business success and munificence, the Ford Foundation would never have been given these opportunities. In other words, but for the tremendous profits Henry Ford generated, Darren Walker might never have been given the chance to question many of those who today actively strive to reprise Ford's business success.

5. For a more extensive discussion on capitalism's variants, strengths, and weaknesses, see William J. Baumol, Robert E. Litan, and Carl J. Schramm, *Good Capitalism, Bad Capitalism and the Economics and Growth and Prosperity* (Newhaven, CT: Yale University Press, 2009).

6. In my considered judgment, much of what may have discolored Pope Francis's views of capitalism comes from his time as a cardinal in Buenos Aires and his front row seat to Argentina's repeated oligarchic foibles. Though Argentina is blessed with abundant natural resources, its political leaders repeatedly stifled the country's economy and failed its people. Argentina has defaulted on its debts nearly a dozen times, more than any other country. Argentines have also suffered repeated bouts of hyperinflation. *Laudato si'*, Pope Francis's powerful encyclical on caring for our common home, the earth, further excoriates all businesses for their neglect of the environment. A quote from *Laudato si'* can be found on the cover of this book. See Francis, *Laudato si'* (Vatican City: Libreria Editrice Vaticana, 2015).

7. Please remember this DJIA discussion when we come to chapter 8, "A Few Words About Indices." How the DJIA came into being and how it has been reconstituted ever since is a story worth telling. It also has significant bearing on how to "beat the market," modern ESG investing, and optimal capital allocation.

8. Dominic Barton, James Manyika, Timothy Koller, Robert Palter, Jonathan Godsall, and Joshua Zoffer, *Measuring the Economic Impact of Short-Termism*, Discussion Paper (Washington, DC: McKinsey Global Institute, February 2017).

9. We return to this topic in chapter 10, "Hardwiring Corporate Goodness."

10. The asset management giant Fidelity is just one of many compelling examples of privately held companies that have thrived over decades, taking some lean years to refresh and reinvent themselves strategically only to emerge better positioned for the longer term. The engineering and construction giant Bechtel and the health care equipment distributor Medline are two others. See chapter 13, "Let's Speak Privately," and chapter 15, "Civics Lessons," for more examples highlighting the principle of comparative advantage, as well as valuable nonprofit initiatives and strategies that promote more inclusive, sustainable growth.

3. ACTIVISTS, THEIR ARGUMENTS—AND A LITTLE ENGINE THAT COULD

1. Lee Ray, Catharina Hillenbrand von der Neyen, Durand D'souza, Valeria Ehrenheim, Lily Chau, Nicolás Gonzalez, Lorenzo Sani, and Aurore Le Galiot, *Do Not Revive Coal: Planned Asia Coal Plants a Danger to Paris* (London: Carbon Tracker, June 2021).

2. For more complete analytics, see David Blitz and Laurens Swinkels, "Does Excluding Sin Stocks Cost Performance?" *SSRN* (June 25, 2021), http://dx.doi.org/10.2139/ssrn.3839065; and Lisa Smith, "Socially Responsible Investing Vs. Sin Stocks," Investopedia, May 12, 2020. In chapter 12, "Crowded Trades," I examine whether ESG indices are creating the same return distortions as sin exclusions. If it turns out they are, some of the world's most sophisticated investors are likely to reject ESG indices at the same time that regulators push hard for their broader adoption.

3. See Shira Tarrant, *The Pornography Industry: What Everyone Needs to Know* (New York: Oxford University Press, 2016). One of Tarrant's more surprising revelations is the role played by JPMorgan Chase, Fortress Investment Group, and the endowment fund of Cornell University in consolidating multiple pornography providers into a single privately held Canadian corporation called MindGeek. MindGeek's primary owner, Bernard Bergemar, recently had an estimated net worth of between $1.2 billion and $2 billion.

4. Bradford Cornell and Aswath Damodaran, "Valuing ESG: Doing Good or Sounding Good?" *SSRN* (March 20, 2020), http://dx.doi.org/10.2139/ssrn.3557432.

5. The environmental commitments of the Charter for Compassion can be found at https://charterforcompassion.org/charter.

4. C-SUITE INSURRECTIONISTS

1. For those unfamiliar, a "Road to Damascus moment" refers to the famous conversion of the Pharisee Saul of Tarsus, who became Saint Paul the Apostle. The phrase is used colloquially to describe a 180-degree change in one's beliefs. As the story is told, Saul brutally persecuted the earliest Christians until a day in 34 CE, when "a light from heaven flashed around him" while he was en route to Damascus on horseback. Saul was blinded and fell to the ground, upon which Jesus of Nazareth is said to have appeared to Saul, moving him to convert to Christianity. Thenceforth, Paul the Apostle became one of the faith's greatest protectors and advocates, forming multiple Christian communities across Asia Minor and Europe and authoring many instructive epistles. Today, Paul's thirteen letters to the local churches he helped found serve as the basis for a significant portion of the Bible known as the New Testament. Here, I am referring to a possible shift in CEO sentiment—from a focus on the short term and shareholders only, to a focus on the long term and all stakeholders, including shareholders.

2. Alan Murray, "America's CEOs Seek a New Purpose for the Corporation," *Fortune*, August 19, 2019.

3. These excerpts are from Lucian Bebchuk and Roberto Tallarita, "The Illusory Promise of Stakeholder Governance," *Cornell Law Review* 106 (December 2020): 91–178. The authors' conclusions are supported by proprietary research conducted into the signatories of the 2019 restatement. Bebchuk and Tallarita learned that only one CEO in their survey had bothered to consult their board before signing it. They cite this as proof that participation did not represent a significant departure from past practice. Had it been significant, prior board approvals would have been necessary.

4. Andrew Winston, "Is the Business Roundtable Statement Just Empty Rhetoric?" *Harvard Business Review*, August 30, 2019.

5. George Magnus, *The Age of Aging: How Demographics Are Changing the Global Economy and Our World* (Singapore: Wiley, 2009). In this book, Magnus presciently

explores every social and commercial aspect relating to humanity's extraordinary, unprecedented demographic shift.

6. There is absolutely no reason for the United States to suffer from the same aging demographics as Europe, Russia, or China. Sensible immigration and family support laws could keep the U.S. population expanding in a healthy manner for decades. Matt Yglesias does a commendable job in stating not only how such laws might work but also why in *One Billion Americans: The Case for Thinking Bigger* (New York: Penguin, 2020).

7. Alicia H. Munnell, Anqi Chen, and Robert L. Siliciano, *The National Retirement Risk Index: An Update from the 2019 SCF* (Boston: Center for Retirement Research at Boston College, January 2021).

8. It is far too easy to see how many countries will experience what Greece went through between 2010 and 2012, when federal insolvency decimated their public pension system and ruined many lives. Retirees with no other sources of income were suddenly forced to fend for themselves, and Greek suicide rates spiked more than 35 percent. For example, at the age of seventy-seven, Dimitris Christoulus announced, "I am not committing suicide; they are killing me," before shooting himself in front of the Parliament in Athens. The global data reported here come from Mercer, *Mercer CFA Institute Global Pension Index 2021: Pension Reform in Challenging Times* (New York: Mercer, 2021).

9. As this book was going to press, BlackRock had fulfilled this promise, publishing MSCI's "implied temperature rise" (ITR) metric for most of their index mutual funds and ETFs. It won't surprise you to learn that temperature alignment in funds turns out to be a key metric for winning new clients. We will return to this in chapter 11, "Inside the ESG Arms Race."

10. As briefly mentioned earlier, scope 1, 2, and 3 emissions are designations that categorize the nature and source of one's greenhouse gas footprint. Scope 1 includes all *direct emissions* under one's control, such as those from heating, lighting, and fleet vehicles. Scope 2 includes *indirect emissions*, including those from one's electricity and other power providers. For scope 2 emissions, information must be provided on how the providers generate their energy. Scope 3 includes *all other indirect emissions*, including those of one's suppliers. These include emissions associated with things like business travel, supply chains, waste, and water use. Scope 3 emissions are by far the largest part of most businesses' carbon footprints and are of course the hardest type to measure and manage.

5. WHAT IF +1°C = −$100 TRILLION?

1. If you have not already done so, check out Mark Lynas's incremental analysis of what life on Earth might be like with every rising degree. Mark Lynas, *Our Final Warning: Six Degrees of Climate Emergency* (London: HarperCollins, 2020). As the *Times* of London wrote, Lynas's is "the clearest account we have come across of what climate change will look like, depending on what we do about it."

2. BlackRock Investment Institute, *Climate Aware Capital Market Assumptions* (New York: BlackRock, March 2021).

3. BlackRock, *Climate Aware Capital Market Assumptions*.

4. BlackRock, *Climate Aware Capital Market Assumptions*.

6. WHAT'S THE UNITED NATIONS GOT TO DO WITH IT?

1. Francis Fukuyama, *The End of History and the Last Man* (New York: Free Press, 1992).
2. By "modern" I mean over the past seventy-five years, not as defined by the University of Oxford, whose designation includes everything that follows Diocletian and the fall of the Roman Empire.
3. Today the UN General Assembly has 193 members, including the Vatican.
4. The distinguished Harvard Law School professor (and, I must add, supremely eloquent Eleanor Roosevelt biographer!), Mary Ann Glendon, has persuasively argued that the Cold War effectively tore the declaration in half, with the United States publicly championing its political and civil rights components, while the Soviet Union championed its social and economic provisions. In the end, of course, neither side can lay claim to victory even on the limited grounds they chose. The United States has yet to achieve full civil rights, and the proven inability of any command economy to generate widespread, equal economic and financial abundance has been exhaustively documented. After their atrocities in the Ukraine, Russia no longer has standing for any principle, moral or financial. Like Eleanor Roosevelt herself, Professor Glendon advocates core principles as the basis for lasting progress.
5. While much is made of today's inequality *within* countries, global Gini coefficients—a measure of inequality *between* nations—have demonstrably improved over the past four decades. Stated plainly, poorer countries have been catching up to richer countries since 1980. This achievement is often overlooked. The largest personal fortunes ever assembled are not those of Bezos, Buffett, Gates, or any monarch like Queen Elizabeth II, moreover. Jakob Fugger of Germany (1459–1525) and John D. Rockefeller (1839–1937) are estimated to have been the wealthiest humans in history, with personal fortunes exceeding $400 billion, once adjusted for inflation. Modern Gini coefficients within the United States, Europe, and China nevertheless suggest more wealth is concentrated in the top 1 percent today than at any other time on record, including during the eras of Fugger and Rockefeller. The case for improving many national Gini coefficients, such as that argued by Thomas Piketty, is compelling.
6. Please allow me to share a personal Kofi Annan story. I was flying from Zurich back to New York in 2009, following an international conference. I had used my excess air miles to upgrade to first class on Swiss Air. Only one other person was with me: the UN secretary-general himself. After eating lunch separately, Mr. Annan came over to introduce himself and ask what I was reading. It was a book on Swiss armed neutrality during the Second World War by Stephen Halbrook. "Switzerland was in a far more difficult position than most realize during World War II," Mr. Annan shared. I finished the book before landing and inscribed it to the secretary-general with my best wishes. Three weeks later, I received a handwritten note from the secretary-general of the United Nations, thanking me profusely for my kindness. "It was even better than I expected," he wrote. I experienced what many had come to know: Kofi Annan was a deeply thoughtful, caring, gracious international diplomat of rare skill and—for me and many others—a personal encounter with eternal impact.
7. PRI, *PRI Board Report: 2020 Signatory General Meeting* (London: PRI, 2020).
8. "What Are the Principles for Responsible Investment?" PRI, https://www.unpri.org/about-us/what-are-the-principles-for-responsible-investment.

7. MATERIALITY

1. As this book was being written, Facebook changed its name to Meta. The episodes described in this chapter occurred while it was still Facebook. For consistency, for the balance of this book, it will remain Facebook.
2. The largest corporate fine in U.S. history was levied on BP for the Deepwater Horizon oil spill fiasco, a $20.8 billion whammy. Bank of America, Volkswagen, JPMorgan Chase, and BNP Paribas round out the top five, with fines ranging from $8 billion to $18 billion for activities that included mortgage fraud, doctored emissions data, money laundering, and other less specific "criminal patterns" of behavior.
3. The data on innovation gaps between the European Union and the United States speaks for itself. Even though the European Union and the United States have roughly comparable global GDP shares, the United States generates three times more equity "unicorns" per year than does the European Union and nearly half of the world's total. This is why there are no European equivalents of Apple, Facebook, Google, Microsoft, or Twitter. China is second after the United States in terms of equity "unicorns," with a 26 percent global share. Europe needs to innovate more.
4. The decision by Royal Dutch Shell (now Shell, PLC) to drop its Dutch listing and incorporate solely in the United Kingdom is but the most recent example of a major corporation moving to a more favorable jurisdiction. An extensive and deep literature on the so-called race-in-laxity or race-to-the-bottom phenomenon exists that will not be explored here. It draws from Adolf A. Berle's book *The Modern Corporation and Private Property* (Piscataway, NJ: Transaction, 1932), as well as multiple Supreme Court cases, including Justice Louis Brandeis's important 1933 decision, *Liggett Co. v. Lee* (288 U.S. 517, 558–59). Whether every move to more accommodative regulatory standards is a "race to the bottom" or "a race for greater efficiency" makes for great debate. Personally, I favor competition between jurisdictions no less than between corporations and industries. Let Schumpeter be Schumpeter!
5. Dane Christensen, George Serafeim and Anywhere Sikochi, "Why Is Corporate Virtue in the Eye of the Beholder? The Case of ESG Ratings," *Accounting Review* 97, no. 1 (January 2022): 147–75.
6. By the time this book went to print, MSCI had changed their relative rankings of Coke and Pepsi. I asked to publish their earlier analysis, but MSCI did not grant me permission to do so.
7. Florian Berg, Julian F. Kölbel, and Roberto Rigobon, "Aggregate Confusion: The Divergence of ESG Ratings," *SSRN* (August 15, 2019), http://dx.doi.org/10.2139/ssrn.3438533.
8. R. Boffo and R. Patalano, *ESG Investing: Practices, Progress and Challenges* (Paris: OECD, 2020), https://www.oecd.org/finance/ESG-Investing-Practices-Progress-Challenges .pdf.
9. Jay Clayton, "Remarks at the Meeting of the Asset Management Advisory Committee," U.S. Securities and Exchange Committee, public hearing, May 27, 2020.

8. A FEW WORDS ABOUT INDICES

1. Fictions despised? Journalism has sure changed. . . .
2. The legendary investor Benjamin Graham owes a special debt of gratitude to Charles Dow. Without Dow's index innovation, Graham's iconic book *The Intelligent Investor*

would never have conceived of "Mr. Market." One of Graham's more memorable observations about Mr. Market is that he is a voting machine in the short run and a weighing machine in the long run. How so? In the short run, Graham notes, Mr. Market can be irrational, euphoric, and moody—so he's a voting machine. But longer term, Mr. Market invariably reflects broad economic trends, corporate earnings, and distributions. This makes him a weighing machine. Graham argues the most important attribute of successful investment is patience. Intelligent investors buy stocks when Mr. Market is in a pessimistic mood and trim positions when he's become too giddy. Given how economies expand over time, though, you must own stocks over long periods of time to realize their full potential. Warren Buffett described *The Intelligent Investor* as "by far the best book on investing ever written." Graham's masterpiece doubtless led Buffett to coin his most quoted piece of investment wisdom: "Be greedy when the market is fearful, and fearful when the market is greedy." See Benjamin Graham, *The Intelligent Investor* (New York: HarperCollins, 1973).

3. GE was added to the DJIA in 1896 and removed in 2018, making for a record-breaking 122-year run. CEO Jeff Immelt's seventeen-year tenure, from 2001 to 2018, failed to match Jack Welch's remarkable record of shareholder returns. GE outperformed the S&P 500 by two times on Welch's watch. While Welch is credited with managerial gifts, his tenure at GE was also defined by more than six hundred strategic acquisitions. GE's stock attained its all-time, stock-split-adjusted high of $55.57 on September 29, 2000—and fell by 90 percent exactly twenty years later. Over this same period, the S&P 500 rose 281 percent. Schumpeter again? Yes, but also poor management. Corporate leadership is a crucial determinant of corporate success. Immelt failed where Welch shone.

4. The Panic of 1896 was precipitated by a large drop in national silver reserves and the concerns this raised for gold prices and deflation. Because the Fed now effectively prints money at will, we forget that the United States once had a very different approach to managing its money supply. Like other nations, we went back and forth on the use of a "metallic" currency standard—with the U.S. dollar pegged at different times to gold, silver, or a combination of gold and silver. A drop in official precious metal reserves required reducing currency in circulation, invariably leading to tighter monetary conditions, higher rates, and bank failures. Official gold and silver reserves effectively determined how high interest rates needed to go. Keynes once referred to the gold standard as a "barbarous relic." During the election of 1896, the Populist Party candidate William Jennings Bryan implored central bankers and other political leaders "not to crucify mankind upon a cross of gold." Today, of course, central banks no longer use metallic standards to set their interest rates or notes in circulation, with the Swiss National Bank the last to end its gold link to the Swiss franc in 1999. Instead, central banks are free to expand and contract their balance sheets as they see fit, consistent with financial and price stability. In addition to these two objectives, the U.S. central bank, the Federal Reserve, is further obligated to pursue full employment. Whether it will prove possible for one institution to achieve all these objectives reliably is very much a work in progress.

5. Keep an eye out for the CRSP US Total Market Index. It strives to represent 100 percent of the investable U.S. stock market, including large-, mid-, small-, and micro-cap stocks. It is also the basis for the largest ETF in the world: Vanguard's VTI.

6. The factors of quality, growth, value, and momentum explain as much as 95 percent of equity and bond price movements. Successfully tilting and timing these factors in concert with the economic cycle generates extraordinary excess return, or alpha.

9. VALUES VERSUS VALUATIONS

1. Richard Scott Stultz, "An Examination of the Efficacy of Christian-Based Socially Responsible Mutual Funds," *Journal of Impact and ESG Investing* 1, no. 2 (Winter 2020): 105–19.
2. Jan-Carl Plagge and Douglas M. Grim, "Have Investors Paid a Performance Price? Examining the Behavior of ESG Equity Funds," *Journal of Portfolio Management* 46, no. 3 (February 2020): 123–40.
3. Tensie Whelan, Ulrich Atz, Tracy Van Holt, and Casey Clark, *ESG and Financial Performance: Uncovering the Relationship by Aggregating Evidence from 1,000 Plus Studies Published Between 2015–2020* (New York: NYU Stern Center for Sustainable Business, February 10, 2021).
4. Lauren Solberg, "Why Sustainable Strategies Outperformed in 2021," Morningstar, January 19, 2022, https://www.morningstar.com/articles/1075190/why-sustainable -strategies-outperformed-in-2021.
5. The best performing stock in the S&P 500 during 2021 was Devon Energy: up 185 percent, more than seven times the broader market. It's not hard to see how S&P 500 index funds screened to omit firms like Devon Energy would underperform.
6. Shane Shifflett, "Funds Go Green, but Sometimes in Name Only," *Wall Street Journal*, September 9, 2021.
7. "Sustainable Investing: A 'Why Not' Moment," BlackRock Investment Institute, May 9, 2018, https://www.blackrock.com/us/individual/insights/blackrock-investment-institute /sustainable-investing-is-the-answer.
8. As we will see in the next few chapters, most asset managers claim that ESG considerations could provide considerable alpha potential. Such a claim is unquestionably true. But a recent Scientific Beta project vigorously rejects the additionality of most ESG metrics, arguing ESG ratings add no value over information already provided by sector classifications or well-known factor attributes. According to their research, "ESG strategies perform like simple quality strategies mechanically constructed from accounting ratios." Time will bring better data and more forecastable risks. Until such time, other considerations will need to come into play. See Giovanni Bruno, Mikheil Esakia, and Felix Goltz, *"Honey, I Shrunk the ESG Alpha": Risk-Adjusting ESG Portfolio Returns* (Singapore: Scientific Beta, April 2021).
9. Øystein Olsen and Trond Grande, "Climate Risk in the Government Pension Fund Global," letter to the Ministry of Finance (Oslo: Norges Bank, July 2, 2021).
10. Norges Bank Investment Management, *Responsible Investment: Government Pension Fund Global*, no. 7 (Oslo: Norges Bank Investment Management, 2020).
11. Olsen and Grande, "Climate Risk," 14–15. Note that a parallel report by an expert panel expressed strong views on using the fund's stewardship practices more aggressively, demanding corporate stress tests of all climate outcomes with the goal of reducing risks.
12. Ellen Quigley, Emily Bugden, and Anthony Odgers, *Divestment: Advantages and Disadvantages for the University of Cambridge*, Paper no. 20.09.21.SM6 (Cambridge: University of Cambridge, September 2020).
13. Norwegian Ministry of Finance, press release no. 44/2006, June 2006. For a comprehensive overview of this decision and the dynamics behind it, see Andrew Ang, *The Norwegian Government Pension Fund: The Divestiture of Wal-Mart Stores Inc.*, Columbia CaseWorks, case ID 080301, Spring 2008, https://www8.gsb.columbia.edu /caseworks/node/256.

10. HARDWIRING CORPORATE GOODNESS

1. Ethan C. Rouen and Charles C. Y. Wang, "Measuring Impact at Just Capital," Harvard Business School Case 119-092, March 2019 (revised April 19, 2019).
2. Specifically, GameChangers sign a nonverifiable pledge to abide by the following principles: (1) Doing good is good for business; (2) leading with purpose improves performance; (3) long-term profitability is generated by incentivizing long-term decisions; (4) people should be treated as human beings, not human resources; (5) the environment and all living species should be protected and cherished; (6) leaders should be supported and rewarded for prioritizing purpose-driven endeavors; (7) customers, suppliers, investors and all stakeholders should be treated with dignity, empathy, and love; (8) communities in which businesses operate should be positively impacted by their presence; (9) people should have a fair stake in the rewards of their work, as well as a voice in decisions that affect them; and (10) interdependence compels business to join together to bring forth a sustainable and thriving world, for all life and for future generations.
3. Michael E. Porter and Mark R. Kramer, "Creating Shared Value: How to Reinvent Capitalism—and Unleash a Wave of Innovation and Growth," *Harvard Business Review*, January–February 2011.
4. George Serafeim, "Social-Impact Efforts That Create Real Value: They Must Be Woven Into Your Strategy and Differentiate Your Company," *Harvard Business Review*, September–October 2020.
5. Alex Edmans, *Grow the Pie: How Great Companies Deliver Both Purpose and Profit* (Cambridge: Cambridge University Press, 2020), 3.

11. INSIDE THE ESG ARMS RACE

1. Clearly, "explicit" will be more demanding than "integrated." How much more demanding is the question. Bloomberg Intelligence, "ESG Assets May Hit $53 Trillion by 2025, A Third of Global AUM," Bloomberg Professional Services, February 23, 2021, https://www.bloomberg.com/professional/blog/esg-assets-may-hit-53-trillion-by-2025-a-third-of-global-aum/.
2. Patricia Kowsmann and Ken Brown, "Fired Executive Says Deutsche Bank's DWS Overstated Sustainable-Investing Efforts," *Wall Street Journal*, August 1, 2021.
3. Adeline Diab and Maxime Boucher, "ESG ETF Watch: Are ESG ETFs Doing What They Claim?" Bloomberg Intelligence, September 29, 2021.
4. Alyssa Buttermark, J. Tyler Cloherty, and Benjamin F. Phillips, *It's Not Easy Being Green: Managing Authentic Transformation Within Sustainable Investing* (New York: Casey Quirk, 2021).
5. Joseph Chi, Eric Geffray, Jeromey Thornton, and Jim Whittington, "ESG Data, Ratings, and Investor Objectives," Dimensional Fund Advisors, May 17, 2021.
6. Brian Bedell and Debbie Jones, "Are We at an Inflection Point in Asset Manager ESG Product Launches?" #dbSustainability Tracker, August 31, 2021.
7. As one would expect, several sectors are more heavily weighted in the S&P Net Zero 2050 Paris-Aligned ESG Index versus the S&P 500, including information technology, health care, communication services, and financials. Taken together, they recently composed 85 percent of the Paris-aligned index versus only 60 percent of the

unaligned index. This heavy reliance on tech and health care in most ESG funds raises issues about crowded trades, the topic of our next chapter.

8. InfluenceMap, *Climate Funds: Are They Paris Aligned?* (London: InfluenceMap, August 27, 2021).

9. Professor Scott Galloway of NYU Stern stated this concern more colorfully on his blog. ESG "capital flows will produce a bevy of firms that tap into the consensual hallucination that society's biggest problems can be fixed as our stock portfolios explode in value. Yeah . . . wouldn't that be nice. No, just as turning back Hitler and smallpox took leadership, treasure, and blood, so will this." Scott Galloway, "Jumping the SPAC," *No Mercy/No Malice* (blog), September 17, 2021. However, Galloway should also acknowledge that real shareholder value can and often is created by greater corporate mindfulness of ESG concerns. When companies strive for noble purposes and become more attentive to their employees, suppliers, and communities, they can and often do benefit their shareholders as well.

10. NN Investment Partners, *Green Bond Funds Impact Report 2020: Financing the Transition to a More Sustainable World* (The Hague: NN Investment Partners, 2020).

11. BlackRock, *Sustainable Investing in Fixed Income: Our Proprietary PEXT/NEXT Framework* (New York: BlackRock, September 2021).

12. Hortense Bioy, Elizabeth Stuart, and Andy Pettit, *SFDR—The First 20 Days: What the Early Batch of New Disclosures Are Telling Us So Far* (Morningstar, March 30, 2021).

13. U.S. Securities and Exchange Commission Asset Management Advisory Committee, *Recommendations for ESG* (Washington, DC: U.S. Securities and Exchange Commission, July 7, 2021).

14. Investment Company Institute, *Funds' Use of ESG Integration and Sustainable Investing Strategies: An Introduction* (Washington, DC: Investment Company Institute, July 2020).

15. Jennifer Choi to Raluca Tircoci-Craciun, "Re: Public Comment on *Recommendations on Sustainability-Related Practices, Policies, Procedures and Disclosure in Asset Management*," ICI Global, August 13, 2021, https://www.ici.org/system/files/2021-08/ioscoltr.pdf.

12. CROWDED TRADES

1. It is not my intention here to delve into the compelling debate on when, where, and how the efficient market hypothesis breaks down. It's worth noting it was defended by the Princeton economist Burton G. Malkiel in a widely cited article. See Burton G. Malkiel, "The Efficient Market Hypothesis and Its Critics," *Journal of Economic Perspectives* 17, no. 1 (Winter 2003): 59–82. Eugene Fama and his fellow University of Chicago professor Kenneth French went on to develop additional models that built upon their original hypothesis. These models ultimately predicted about 90 percent of all asset price movements, an astounding achievement. Other theorists and practitioners have taken these studies even further, showing that certain factors like momentum, quality, growth, and minimum volatility have significant explanatory powers for stock and bond prices.

2. Today, of course, Sir John's legacy is much greater than the billions he earned. The John Templeton Foundation and its related entities support a vision "of infinite scientific and spiritual progress, in which all people aspire to and attain a deeper understanding

of the universe and their place in it." To what end? "We look forward to a world where people are curious about the wonders of the universe, motivated to pursue lives of meaning and purpose, and overwhelmed by great and selfless love." The John Templeton Foundation and its affiliates are exemplars of hope.

3. Savita Subramanian, Marisa Sullivan, and Panos Seretis, *ESG Matters: ESG 30 Percent of Equity Inflows YTD* (New York: Bank of America Securities, August 5, 2021).

4. Obviously, idiosyncratic risks skew positively and negatively. For example, Morningstar highlights how two tech names—Nvidia and Microsoft—accounted for a large portion of their sustainable index outperformance in 2021. Nvidia returned 125 percent in 2021, while Microsoft gained 52 percent. Nvidia has a weight three times greater in their U.S. Sustainability Leaders Index, and Microsoft two times greater, versus their broader index. By comparison, Exxon—which was up 57 percent in 2021—has a weight of less than 1 percent in Morningstar's broad index, meaning its outperformance did little to harm its performance measurement. Remember what I told you in chapter 8? *Index providers today are more powerful than popes or presidents!* They get to skew performance measurement in ways that make their success more probable, and often do.

5. Jean-Marie Dumas and Jean-Jacques Barberis, *Temperature Scores: An Innovative Tool for ESG Fundamental Investors*, Investment Insights Blue Paper (Paris: Amundi Asset Management, April 2021).

6. Jennifer Choi to Raluca Tircoci-Craciun, "Re: Public Comment on *Recommendations on Sustainability-Related Practices, Policies, Procedures and Disclosure in Asset Management*," ICI Global, August 13, 2021, https://www.ici.org/system/files/2021-08 /ioscoltr.pdf.

13. LET'S SPEAK PRIVATELY

1. John Asker, Joan Farre-Mensa, and Alexander Ljungqvist, "Corporate Investment and Stock Market Listing: A Puzzle?" *Review of Financial Studies* 28, no. 2 (February 2015): 342–90.

2. Publix Super Markets, H-E-B, C&S Wholesale Grocers, and Meijer have more employees than the publicly listed Kroger. "America's Largest Private Companies," *Forbes*, September 2021.

3. Tim Dickinson, "Inside the Koch Brothers' Toxic Empire," *Rolling Stone*, September 24, 2014.

4. Anjli Raval, "A $140 Billion Asset Sale: The Investors Cashing In on Big Oil's Push to Net Zero," *Financial Times*, July 6, 2021.

5. Raval, "A $140 Billion Asset Sale."

6. OECD, *The Risk of Corruption In and Around State-Owned Enterprises: What Do We Know?* (Paris: OECD, August 27, 2018).

14. FIGHT OR FLEE?

1. Eleonora Broccardo, Oliver Hart, and Luigi Zingales, *Exit vs. Voice*, European Corporate Governance Institute—Finance Working Paper No. 694/2020 (Brussels: ECGI, August 2020).

2. Engine No. 1 and Witold J. Henisz, *A New Way of Seeing Value: Introducing the Engine No. 1 Total Value Framework* (San Francisco: Engine No. 1, September 2021).

3. John C. Coates, *The Future of Corporate Governance Part I: The Problem of Twelve* (Cambridge, MA: Harvard Law School, September 20, 2018).

4. State Street Global Advisors, *Stewardship Report 2020* (Boston: State Street, 2021).

5. BlackRock Investment Stewardship, *Our 2021 Stewardship Expectations* (New York: Blackrock, 2020).

6. See, for example, Rob Berridge, "With New Power Comes New Responsibility: How Asset Managers Can Improve Their Voting Record on Climate in 2021," Ceres, December 22, 2020; and Majority Action, *Climate in the Boardroom: How Asset Manager Voting Shaped Corporate Climate Action in 2021* (San Francisco: Majority Action, 2021).

7. In 2016/17, the leading proxy advisory firm ISS supported 71 percent of shareholder proposals versus only 15 percent for Vanguard, 16 percent for BlackRock, and 33 percent for State Street. Fidelity was even lower, at 14 percent. Source: ISS Analytics.

8. John Rekenthaler, "Three Solutions for Index-Fund Voting," Morningstar, June 14, 2021, https://www.morningstar.com/articles/1042975/3-solutions-for-index-fund-voting.

9. John C. Bogle, "Bogle Sounds a Warning on Index Funds," *Wall Street Journal*, November 29, 2018.

10. In particular, BlackRock's move to introduce choice in stewardship should mute the criticisms of folks like Stephen Soukup, the author of *The Dictatorship of Woke Capital: How Political Correctness Captured Big Business* (New York: Encounter, 2021), as well as the suggestion by Charlie Munger that CEO Larry Fink has somehow become an "emperor." Choice and dictatorship are antithetical.

15. CIVICS LESSONS

1. To nominate an organization to be an exemplar of hope, please email terry@1pointsix.com or visit www.1pointsix.com. Following a formal vetting process, all qualified institutions will be added to the Exemplars of Hope website. Note all nominated organizations must verifiably promote economic inclusivity and/or environmental sustainability in a scalable manner. Importantly, their methodologies must not penalize economic growth, inclusivity, or sustainability in any way; all three objectives must be supported by their specific pedagogies.

2. Raghuram Rajan, *The Third Pillar: How Markets and the State Leave Community Behind* (New York: Penguin, 2019), xvii.

3. Rajan, *The Third Pillar*, 266–67.

4. Charities Aid Foundation, *CAF World Giving Index: Ten Years of Giving Trends* (London: CAF, October 2019).

5. Rajan, *The Third Pillar*, 248.

6. John F. Helliwell, Richard Layard, Jeffrey D. Sachs, Jan-Emmanuel De Neve, Lara B. Aknin, and Shun Wang, eds., *World Happiness Report 2021* (New York: Sustainable Development Solutions Network, 2021).

7. Raj Chetty, Nathaniel Hendren, Patrick Kline, and Emmanuel Saez, "Where Is the Land of Opportunity? The Geography of Intergenerational Mobility in the United States," *Quarterly Journal of Economics* 129, no. 4 (November 2014): 1553–623.

8. Church of Jesus Christ of Latter-day Saints, *Providing in the Lord's Way: A Leader's Guide to Welfare* (Salt Lake City, UT: Corporation of the Presidents of the Church of Jesus Christ of Latter-day Saints, 1990).

9. In fact, the LDS approach to self-reliance and communal responsibility has multiple parallels with Catholic Social Teaching (CST). Two principles compose CST's framework: *subsidiarity* and *solidarity*. Subsidiarity means that all social issues must be dealt with at the most proximate level consistent with their full resolution. As with LDS, this begins with individual responsibility. Like LDS, the Catholic Church maintains that everyone is responsible for their own behavior; if they weren't, free will and earned salvation would have no meaning. That said, if an individual cannot succeed on their own, CST says the next level of responsibility falls upon their family or their extended family. Where family is not able to help sufficiently, the Church and community are called upon to provide additional support. Solidarity means that, no matter what happens, we are always and everywhere our brothers' and sisters' keepers. Human souls are not divided by income, race, national boundaries, or separate faiths. Every human soul matters equally. By implication, CST believes no one should rest until everyone can rest. When you put subsidiarity and solidarity together, bonds of personal and mutual responsibility become inextricably interwoven. Under CST, communities must come together to solve communal problems—and states must come together to solve global problems. CST is demanding. It requires work that is never-ending. In this respect, it is also simultaneously endlessly rewarding.

10. Jeremy Beer, *The Philanthropic Revolution: An Alternative History of American Charity* (Philadelphia: University of Pennsylvania Press, 2015).

16. IMPACT INVESTING AT SCALE

1. According to publicly available data, since inception through December 2021, Baillie Gifford's Positive Change Fund generated 18.87 percent of excess annual return versus their MSCI benchmark. Other public equity impact managers who generated some positive returns over this period include teams at Aberdeen, Arjuna Capital, Black-Rock, BMO, Impax Asset Management, M&G Investments, NN Investment Partners, Wellington, and WHEB. Doing well and doing good in publicly listed equity strategies is not only possible, it's already happening.

2. I do not mean to suggest that impact measurement is a precise and developed science. It isn't. That said, it is more advanced than many presume. For a detailed description of one approach called IMM, see Chris Addy, Maya Chorengel, Mariah Collins, and Michael Etzel, "Calculating the Value of Impact Investing: An Evidence-Based Way to Estimate Social and Environmental Returns," *Harvard Business Review*, January–February 2019. There are considerable opportunities for further research as well as industry competition in impact investment measurement. Asset owners should seek more transparency from their managers about what impact their capital is having on the social and environmental outcomes they most care about.

3. The Operating Principles for Impact Management were launched in 2019 by a group of global investors in consultation with the International Finance Corporation, a sister organization of the World Bank Group. They are helping to set best-practice standards for measuring and reporting impact in accordance with nine principles, beginning with strategic intent through to origination and structuring, management, and impact at exit. These efforts help promote sanctity and trust in the impact investment field. For further details, visit https://www.impactprinciples.org/9-principles.

4. John Doerr, *Speed and Scale: An Action Plan for Solving Our Climate Crisis Now* (New York: Penguin Random House, 2021). To be clear, I do not agree with Doerr on this

claim. The world has an emissions crisis, not a fossil-fuel crisis. If we can find ways to create clean fossil-fuel energy plants or to remove carbon from the air at scale, fossil fuels would remain a safe and reliable part of our energy mix. Some fossil fuels will undoubtedly be needed. Ultimately, every country must prioritize clean, affordable, safe, and reliable energy systems. In many cases, this will require fossil fuels. It will also require additional nuclear capacity.

5. BlackRock Investment Institute, *The Big Emerging Question: How to Finance the Net-Zero Transition in Emerging Markets* (New York: BlackRock, October 2021).

6. The Global Steering Group for Impact Investment is an independent charity registered in the United Kingdom that catalyzes impact investments and entrepreneurship to benefit people and the planet. It is the successor organization to the Social Impact Investment Taskforce established under the United Kingdom's presidency of the G8 in 2015. For further details, visit https://www.gsgii.org.

7. Ronald Cohen, *Impact: Reshaping Capitalism to Drive Real Change* (London: Ebury, 2020).

17. THE 1.6% "SOLUTION"

1. United Nations, *United Nations Secretary-General's Roadmap for Financing the 2030 Agenda for Sustainable Development: 2019–2021* (United Nations, 2019).

2. BlackRock Investment Institute, *The Big Emerging Question* (New York: BlackRock, October 2021).

3. Coincidentally, the McKinsey Global Institute came up with this same, incremental investment figure of $3.5 trillion in their recent report *The Net Zero Transition: What It Would Cost, What It Could Bring,* January 2022.

4. The data I cite here come from Allianz Research, *Allianz Global Wealth Report 2021* (Munich: Allianz Research, October 7, 2021). This said, other estimates for global investable wealth are as high as $400 trillion because they include broader measures of individual wealth. As we need an incremental $3.5 trillion of impact investing per year to achieve most UN SDGs, whatever denominator one ends up with will be a direct function of one's chosen numerator.

INDEX